A Text Book Of

VLSI WITH VHDL

(22062)
(ELECTIVE)

Semester - VI

THIRD YEAR DIPLOMA IN
ELECTRONICS ENGINEERING GROUPS

As Per MSBTE's I Scheme Syllabus

VIJAY G. YANGALWAR

M.Sc., LL.B., L.M.I.S.T.E.

Retd. HOD, Deptt. of Electronics Engg.

Government Polytechnic,

NAGPUR.

N4571

VLSI WITH VHDL (Sem. VI)   **ISBN 978-93-89825-74-9**

First Edition : **January 2020**

© : **Author**

Published By :
NIRALI PRAKASHAN
Abhyudaya Pragati, 1312, Shivaji Nagar
Off J.M. Road, PUNE – 411005
Tel - (020) 25512336/37/39, Fax - (020) 25511379
Email : niralipune@pragationline.com

➤ DISTRIBUTION CENTRES

PUNE

Nirali Prakashan : 119, Budhwar Peth, Jogeshwari Mandir Lane, Pune 411002, Maharashtra
(For orders within Pune)   Tel : (020) 2445 2044, Fax : (020) 2445 1538; Mobile : 9657703145
Email : bookorder@pragationline.com, niralilocal@pragationline.com

Nirali Prakashan : S. No. 28/27, Dhyari, Near Pari Company, Pune 411041
(For orders outside Pune)   Tel : (020) 24690204 Fax : (020) 24690316; Mobile : 9657703143
Email : dhyari@pragationline.com, bookorder@pragationline.com

MUMBAI

Nirali Prakashan : 385, S.V.P. Road, Rasdhara Co-op. Hsg. Society Ltd.,
Girgaum, Mumbai 400004, Maharashtra; Mobile : 9320129587
Tel : (022) 2385 6339 / 2386 9976, Fax : (022) 2386 9976
Email : niralimumbai@pragationline.com

➤ DISTRIBUTION BRANCHES

JALGAON

Nirali Prakashan : 34, V. V. Golani Market, Navi Peth, Jalgaon 425001,
Maharashtra, Tel : (0257) 222 0395, Mob : 94234 91860
Email : niralijalgoan@pragationline.com

KOLHAPUR

Nirali Prakashan : New Mahadvar Road, Kedar Plaza, 1st Floor Opp. IDBI Bank
Kolhapur 416 012, Maharashtra. Mob : 9850046155
Email : niralikolhapur@pragationline.com

NAGPUR

Nirali Prakashan : Above Maratha Mandir, Shop No. 3, First Floor,
Rani Jhanshi Square, Sitabuldi, Nagpur 440012, Maharashtra
Tel : (0712) 254 7129

DELHI

Nirali Prakashan : 4593/15, Basement, Agarwal Lane, Ansari Road, Daryaganj
Near Times of India Building, New Delhi 110002 Mob : 08505972553
Email : niralidelhi@pragationline.com

BENGALURU

Nirali Prakashan : Maitri Ground Floor, Jaya Apartments, No. 99, 6th Cross, 6th Main,
Malleswaram, Bengaluru 560 003, Karnataka
Mob : +91 9449043034
Email: niralibangalore@pragationline.com

niralipune@pragationline.com | www.pragationline.com

Also find us on www.facebook.com/niralibooks

Preface . . .

I take an opportunity to present the text book entitled **"VLSI with VHDL"** to the students of sixth semester for Electronics Engineering Groups. The main objective of this book is to present the subject matter in a most concise, compact, to the point, simple and lucid manner. The book has been written strictly according to the revised syllabus introduced by MSBTE from June 2019.

The book has been written constantly keeping in mind the requirements of all the students regarding the latest and changing trend of the board (MSBTE) examinations. It is also written in an easy style with full details and illustrations.

Each chapter gives important points, practice questions and questions asked in board examinations. In short, the book is expected to meet the crying needs of Diploma students of Electronics and Communication Engineering Groups because it gives the theoretical and practical knowledge of the VLSI with VHDL language.

The author presents a grateful acknowledgement to various sources, drawn upon in the presentations of this text.

I record my deepest appreciation to my family members for their valuable understanding and co-operation during the spare time which was available tooling to academic, household and social commitment, the book was under preparation. Without their support, patience, love and understanding, I could have not completed this text in a short period of time.

It is my privilege to record indeptness to the publishers Shri. Dineshbhai Furia and Shri. Jignesh Furia for their kind interest in entire work. My thanks are also due to Mr. Rahul Thorat, Mrs. Varsha Bodake and Shri. Kiran Velankar of Nirali Prakashan, Pune for publishing this book in good and presentable manner within very short time.

Although every care has been taken to check mistakes and misprints, yet it is very difficult to claim perfection. Any undetected and unintentional errors, omissions, suggestions etc. from students, colleagues, teachers and practicizing engineers for improvements brought to are notice in good spirit are most welcome.

Vijay G. Yangalwar

Syllabus ...

1. Advanced Digital Design and ASIC, FPGA and CPLD

1.1 Review of Sequential Logic : Asynchronous and Synchronous, Metastability, Noise margins, Power fan-out, Skew (Definitions only)

1.2 Moore and Mealy Models, State machine notation.

1.3 Examples on Moore and Mealy : Counter, Sequence detector only.

1.4 ASIC design flow.

1.5 CPLD - Details of internal block diagram.

1.6 FPGA - Architecture, Details of internal block diagram.

2. CMOS Technology Concepts

2.1 Introduction to BJT and CMOS parameters.

2.2 Basic gates using CMOS inverter, NOR, NAND, MOS transistor switches, Transmission gates, CMOS inverter characteristics.

2.3 Complex logic using CMOS.

2.4 Estimation of resistance and capacitance layout.

2.5 Fabrication process : Overview of wafer processing, Oxidation, Epitaxy, Deposition, Ion-implantation and Diffusion, Silicon gate process.

2.6 Basics of NMOS, PMOS and CMOS : nwell, pwell, twin tub process.

3. Introduction to VHDL

3.1 Introduction to HDL : History of VHDL, Pro's and Con's of VHDL.

3.2 VHDL Flow Elements : Entity, Architecture, Configuration, Package, Library only definitions.

3.3 Data types, Operators, Operations.

3.4 Signal, Constant and Variables (Syntax and use).

3.5 VHDL modeling : Data flow, Behavioral, Structural.

4. VHDL Programming

4.1 Concurrent constructs (when, with).

4.2 Sequential constructs (process, if, case, loop, assert, wait).

4.3 VHDL program to implement flip-flop, Counter, Shift register, MUX, DEMUX, ENCODER, DECODER, MOORE, MEALY machines.

4.4 Test bench and its applications.

5. HDL Simulation and Synthesis

5.1 Event scheduling, Sensitivity list, Zero modeling, Simulation cycle, Comparison of software and hardware description language, Delta delay.

5.2 HDL design flow for synthesis.

5.3 Efficient coding styles, Optimizing arithmetic expression, Sharing of complex operator.

☆☆☆

Contents ...

☆☆☆

ADVANCED DIGITAL DESIGN AND ASIC, FPGA AND CPLD

Learning Objectives

(1a) Differentiate between asynchronous and synchronous logic circuit for the given parameters.

(1b) Develop the state diagram, state table for the given sequential logic.

(1c) Develop model of Moore and Mealy machine of the given contents.

(1d) Describe the given ASIC, FPGA and CPLDs.

1.1 SEQUENTIAL LOGIC

1.1.1 Definition

- *A sequential logic is a digital logic in which the value of the output at any time (t) depends on the value of the input at time (t) and on the previous value of input/output.*
- At least one feedback path exists from output to input.
- Memory element is needed to store previous value of input/output.

1.1.2 Requirements

- The requirements of a sequential logic are as under :
    1. A set of statements.
    2. Boolean expression.
    3. Truth table.
    4. Memory elements (FFs).

1.1.3 Characteristics

- The sequential logic has the following characteristics :
    1. It has at least one feedback path from output of the digital system to input of the digital system.
    2. The digital system has a memory capability for storing past information so that the previous input and output values are to be used in determining the previous outputs.

1.1.4 Examples

- The few examples of sequential logic circuits are :
    1. Flip-flops.
    2. Shift registers.
    3. Counters.
    4. Calculators.
    5. Dial combination lock.
    6. Telephone system.

1.1.5 Block Diagram

- A block diagram of a sequential logic circuit is shown in Fig. 1.1.

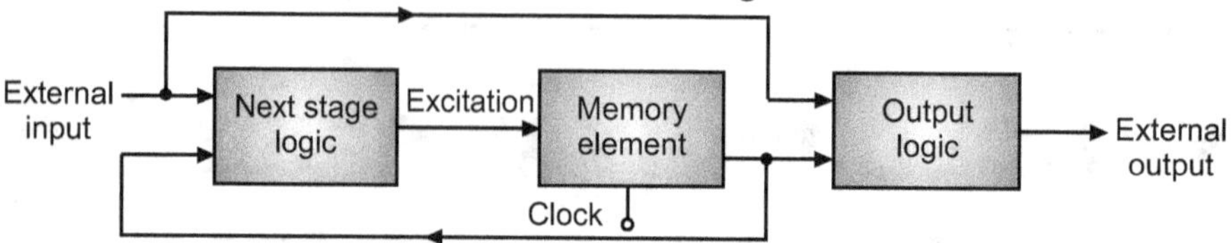

Fig. 1.1 : Block diagram of a sequential circuit

- A sequential circuit consists of next state logic, memory element and output logic.

- Next state logic is a combination circuit, which accepts digital signal from external input and from output of memory element, and generates signal for output logic and inputs to memory element referred to as excitation.

- A memory element is some medium in which one bit of information (1 or 0) can be stored or retained until necessary, and thereafter its contents can be replaced by a new value. The contents of memory element can be changed by the outputs of the combinational circuit, which are connected to its input.

- Output logic is a combinational circuit, which performs some operation on external inputs and memory outputs to generate the external outputs.

- The external output of a sequential circuit depends on external inputs and the present contents of the memory elements. The present contents of the memory element are referred to as **present state** of the memory element (or system). The new contents of the memory elements are referred to as the **next state**. The next state depends on the present state and external inputs.

- Hence, the output of a sequential circuit is a function of the time sequence of inputs and the internal states.

1.1.6 Types

- The sequential circuits are classified into two categories :
  1. Asynchronous sequential circuit.
  2. Synchronous sequential circuit.

1. Asynchronous sequential circuit :

- *A sequential circuit whose behaviour depends upon the sequence in which the input signals change is referred to as Asynchronous sequential **circuit**.*

- The output will be affected whenever the input changes.

- The commonly used memory elements in these circuits are time delay memory devices. These can be regarded as combinational circuits with feedback. They do not require clock pulses.

2. Synchronous sequential circuit :

- *A sequential circuit whose behaviour can be defined from the knowledge of its signal discrete instants of time is referred to as **synchronous sequential circuit**.*

- In these systems, the memory elements are affected only at discrete instant of time.

- The synchronous is achieved by a timing device known as system clock which generates a periodic train of clock pulses.

- The outputs are affected only with the application of a clock pulse.

- This clock signal is the **command signal** that causes the memory element i.e. flip-flop to READ and STORE the code at its input.

- Synchronous circuits have gained considerable domination and wide popularity, and are also known as **clocked sequential circuits**, whereas design of synchronous circuits is more tedious and difficult for their use.

1.2 IMPORTANT PARAMETERS

1.2.1 Metastability

1. Introduction:

- In digital logic circuits, a digital signal is required to be within certain voltage or current limits to represent a '0' or '1' logic level for correct circuit operation; if the signal is within a forbidden intermediate range, it may cause faulty behavior in logic gates, the signal is applied to.

- In metastable states, the circuit may be unable to settle into a stable '0' or '1' logic level within the time required for proper circuit operation. As a result, the circuit can act in unpredictable ways, and may lead to a system failure, sometimes referred to as a glitch.

2. Definition:

- **Metastability is the ability of a digital electronics system to persist for an unbounded time in an unstable equilibrium or metastable state.**

- Metastable state is an excited state of an atom or other system with a longer life time than the other excited states. However, it has a shorter life time than the stable ground state. Atoms in the metastable state remain excited for a considerable time in the order of 10^{-6} to 10^{-3}.

- Metastable states are avoidable in fully synchronous systems, when the input setup and hold time requirements on flip-flops are satisfied.

- **Metastability is a phenomenon that can cause system failure in digital devices, including FPGAs, when a signal is transferred between circuitry in unrelated or asynchronous clock domains.**

3. Example:

- A simple example of metastability can be found in an SR NOR latch, when both Set and Reset inputs are true (R = 1 and S = 1) and then both transition to false (R = 0 and S = 0) at about the same time.

- Both outputs Q and $\overline{Q}$ are initially held at 0 by the simultaneous Set and Reset inputs. After both Set and Reset inputs change to false, the flip flop will (eventually) end up in one of two stable states, one of Q and $\overline{Q}$, true and the other false.

- The final state will depend on which of R or S returns to zero first, chronologically, but if both transition at about the same time, resulting metastability, with intermediate or oscillatory output levels, can take arbitrarily long to resolve to a stable state.

4. Reasons:

    (i) Violation of the set-up or hold times.

    (ii) Clock skews resulting due to different path lengths of the clock pulse provided to the flip-flops.

    (iii) Slower operating frequencies of flip-flops.

    (iv) Asynchronous or rapidly changing states.

5. Advantages:

    (i)   Reducing the clock skew.

    (ii) Using high frequency flip-flops.

    (iii) Avoiding asynchronous inputs.

    (iv) During synchronizer to synchronize asynchronous inputs.

1.2.2 Noise Margin

1. Introduction:

- A digital circuit might be designed to using between 0.0 and 1.2 V, with anything below 0.2 volts considered as '0', and anything above 1.0 V is considered as '1'. Then the noise margin for a '0' would be the amount that a signal is below 0.2 V, and the noise margin for a '1' would be the amount by which a signal exceeds 1.0 V.

- In this case, noise margins are measured as an absolute voltage, not a ratio. Noise margins for CMOS chips are usually much greater than those for TTL because the $V_{OH\,min}$ is closer to the power supply voltage and $V_{OL\,max}$ is closer to zero.
- In practice, noise margins are the amount of noise, that a logic circuit can withstand. Noise margins are generally defined so that positive values ensure proper operation, and negative margins result in compromised operation, or outright failure.

2. **Definition:**

- **Noise margin is the maximum voltage amplitude of extraneous signal that can be algebraically added to the noise-free worst-case input level without causing the output voltage to deviate from the allowable logic voltage level.**
- Noise margin is the ratio by which the signal exceeds the minimum acceptable amount. It is normally measured in decibel (dB).
- Noise margin is the amount by which the signal exceeds the threshold for a proper '0' or '1'.
- **The difference between the operating input voltage level and the input voltage level at which the logic circuit changes from one state to another is known as noise margin.**

1.2.3 Clock Skew

1. **Introduction:**

- The operation of most digital circuits is synchronized by a periodic signal known as a "clock" that dictates the sequence and pacing of the devices on the circuit. The clock is distributed from a single source to all the memory elements of the circuit, which for example could be registers or flip-flops.
- In a circuit using edge-triggered registers, when the clock edge or tick arrives at a register, the register transfers the register input to the register output, and these new output values flow through the **combinational logic** to provide the values at register inputs for the next clock tick.
- Ideally, the input to each memory element reaches its final value in time for the next clock tick so that the behavior of the whole circuit can be predicted exactly.
- The maximum speed at which a system can run must account for the variance that occurs between the various elements of a circuit due to differences in physical composition, temperature, and path length.

2. **Definition:**

- **Clock skew** (sometimes called **timing skew**) is a phenomenon in synchronous digital circuit systems (such as computer system) in which the same sourced clock signal arrives at different components at different times.
- **The instantaneous difference between the readings of any two clocks is called their skew.**

3. **Factors:**

(i) wire-interconnect length,

(ii) temperature variations

(iii) material imperfections, and

(iv) differential input capacitance

- These factors become more critical at high frequency.

4. **Types :**

(i) Positive skew, and

(ii) Negative skew

- Positive skew occurs when the transmitting register receives the clock tick earlier than the receiving register. Negative skew occurs when the receiving register gets the clock tick earlier than the sending register.
- Zero clock skew refers to the arrival of the clock tick simultaneously at transmitting and receiving registers.

1.2.4 Power Dissipation

1. Introduction:

- Dissipation is the result of an irreversible process that takes place in homogeneous thermodynamic system.

- A dissipative process is a process in which energy (internal, bulk flow kinetic, or system potential) is transferred from some initial form.

- The concept of dissipation was introduced in the field of thermodynamics by *William Thomson* (Lord Kelvin) in 1852.

2. Definition:

- **Power dissipation is the maximum power that digital devices such as CMOS can dissipate continuously under the specified thermal conditions.**

- The process in which an electric or electronic device produces heat (other waste energy) as an unwanted byproduct of its primary action.

- **The power required for the logic gate to operate with 50% duty cycle of these specified frequency is known as power dissipation of a logic gate.**

1.2.5 Fan-Out

1. Introduction:

- A perfect logic gate would have **infinite input impedance** and **zero output impedance**, allowing a gate output to drive any number of gate inputs.

- The fan-out is simply the number of inputs that can be connected to an output before the current required by the inputs exceeds the current that can be delivered by the output, while still maintaining correct logic levels.

2. Definition:

- **The fan-out of a logic gate output is the number of gate inputs it can drive.**

- In most designs, logic gates are connected to form more complex circuits, while no logic gate input can be fed by more than one output, it is common for one output to be connected to several inputs.

- The **maximum fan-out** of an output measures its load-driving capability. It is the greatest number of inputs of gates of the same type to which the output can be safely connected.

1.3 FINITE STATE MACHINE (FSM)

1.3.1 Concepts

- A state is a description of the status of a system that is waiting to execute a *transition*. A transition is a set of actions to be executed, when a condition is fulfilled or when an event is received. For example, when using an audio system to listen to the radio, receiving a "*next*" stimulus results in moving to the next station. When the system is in the "*CD*" state, the "*next*" stimulus results in moving to the next track. Identical stimuli trigger different actions depending on the current state.

- In some finite-state machine representations, it is also possible to associate actions with a state:

    1. **an entry action:** performed when entering the state, and

    2. **an exit action:** performed when exiting the state.

1.3.2 Introduction

- The behavior of state machines can be observed in many devices in modern society that perform a predetermined sequence of actions depending on a sequence of events with which they are presented.

- The finite state machine has less computational power than some other models of computation such as the Turing machine.

- The computational power distinction means there are computational tasks that a Turing machine can do but a finite state machine cannot. This is because a finite state machine's memory is limited by the number of states it has.

1.3.3 Definition

- **A finite-state machine (FSM) is a mathematical model of computation.** It is an abstract machine that can be in exactly one of a finite number of states at any given time.

- **A finite state machine is a digital system that traverses through a predetermined sequence of states in an orderly fashion.**

- A finite state machine is a sequential circuit that has some practical bounds governing the member of different conditions.

- A binary counter is an example a finite-state machine as it has 'n' states which are finite micro process or base system as example r is not a finite state machine.

- **A synchronous sequential circuit is called a finite state machine.** It has finite number of states. It is also called a state machine.

- A finite state machine is a model of computation based on a hypothetical machine made of one or more states.

- Only a single state can be active at the same time, so the machine must transition from one state to another in order to perform different actions.

1.3.4 Example

- An example of a simple mechanism that can be modeled by a state machine is a **turnstile**. A turnstile, used to control access to subways and amusement park rides, is a gate with three rotating arms at wrist height, once across the entryway. Initially the arms are locked, blocking the entry, preventing patrons from passing through.

- Depositing a coin or token in a slot on the turnstile unlocks the arms, allowing a single customer to push through. After the customer passes through, the arms are locked again until another coin is inserted.

1.3.5 General Model

- A block diagram of general model of a sequential machine or FSM is shown in Fig. 1.2.

- The function of combinational block is labelled as next state Decoder is to decode the inputs from the outside world, and present state of the machine stored by the memory and to generate as its output, a code called next state code.

- This next state code will become present state, when the memory loads and stores it.

- This process is defined as a STATE CHANGER or a CHANGE OF STATE.

- State changing is a continued process with each new state and the present input conditions being decoded to form the new next state codes.

- Each new succeeding state is a function of the present inputs and past history of these inputs.

- The combinational logic block labelled as the output world code converter has the basic function of decoding the present state of the machine, and present input conditions for the purpose of generating the desired control outputs to the outside world.

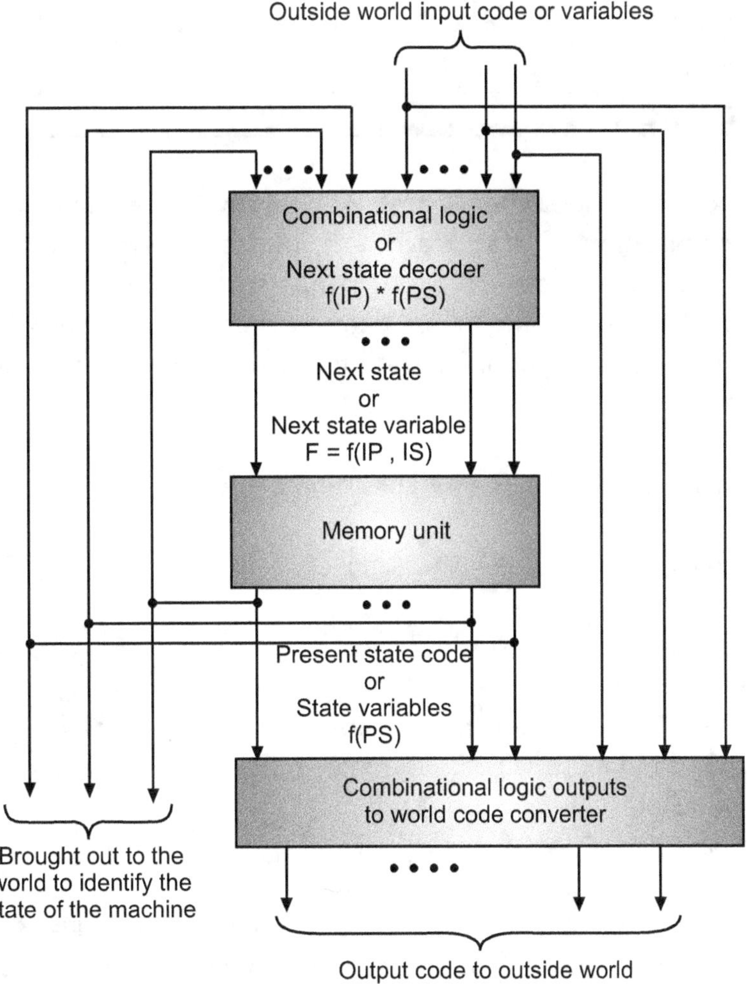

Fig. 1.2 : General model of FSM

1.3.6 Types

- The finite state machines are broadly classified into two categories :
    1. Mealy machine.
    2. Moore machine.

1.3.7 Advantages

1. Flexible
2. Easy to move
3. Low processor overhead
4. Easy determination of reachability of a state

1.3.8 Disadvantages

1. The expected character of deterministic finite state machines cannot be needed in some areas like computer games.
2. The implementation of huge systems using FSM is hard for managing without any idea of design.
3. Not applicable for all domains.
4. The orders of state conversions are inflexible.

1.3.9 Applications

1. Video games,
2. Artificial intelligence,
3. Navigating parsing text,
4. Handling of customers,
5. Handling of network protocols,
6. Software developers,
7. System designers,
8. Vending machines,
9. Traffic lights,
10. Controllers in CPU,

11. Analysis of protocols,   12. Recognition of speech

13. Language processing.

1.4 MEALY STATE MACHINE

1.4.1 Introduction

- The Mealy (state) machine is named after **George H. Mealy**, who presented the concept in 1955.
- In the **theory of computation**, a Mealy machine is a finite-state machine, whose output values are determined both by its current state and the current inputs.
- This is in contrast to a **Moore machine**, whose output values are determined solely by its current state. A **Mealy machine** is a deterministic finite-state transducer for each state and input, at most one transition is possible.

1.4.2 Definition

- A Mealy machine is a 6-tuple $(S, S_0, \Sigma, A, T, G)$ consisting of the following:
    - A finite set of states S.
    - A start state (also called initial state) S_0 which is an element of S
    - A finite set called the input alphabet Σ
    - A finite set called the output alphabet A
    - A transition function $T : S \times \Sigma \rightarrow S$ mapping pairs of a state and an input symbol to the corresponding next state.
    - An output function $G : S \times \Sigma \rightarrow A$ mapping pairs of a state and an input symbol to the corresponding output symbol.

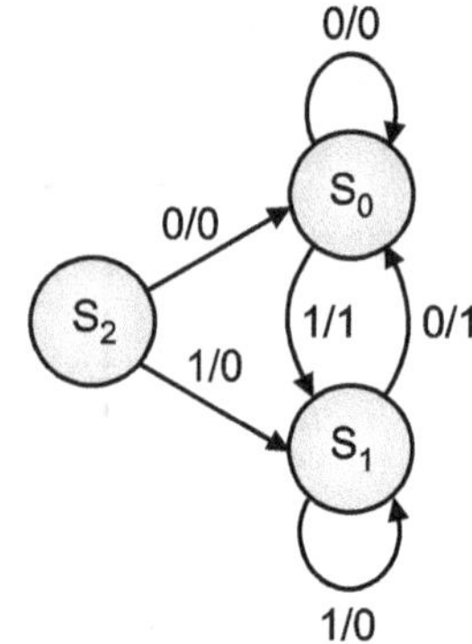

Fig. 1.2 : State diagram of a simple Mealy machine

- **Mealy machine is a machine in which output symbol depends upon the present input symbol and present state of the machine.**
- In the Mealy machine, the output is represented with each input state separated by /.

1.4.3 State Diagram

- The state diagram for a Mealy machine associates an output value with each transition edge, in contrast to the state diagram for a Moore machine, which associates an output value with each state.
- When the input and output alphabet are both Σ, one can also associate to a Mealy Automata in Helix **directed graph** $(S \times \Sigma, (x, i) \rightarrow (T(x, i), G(x, i)))$.
- This graph has as vertices the couples of state and letters, every nodes are of out-degree one, and the successor of (x, i) is the next state of the automata and the letter that the automata output when it is instate x and it reads letter i. This graph is a union of disjoint cycles if the automaton is bireversible.

1. **Simple example:**
- A simple Mealy machine has one input and one output. Each transition edge is labelled with the value of the input (shown in red) and the value of the output (shown in blue).
- The machine starts in state S_1 and the output is the EX-OR of the two most-recent input values; thus, the machine implements an edge detector, outputting a one every time the input flips and a zero otherwise.

2. **Complex example:**
- More complex Mealy machines can have multiple inputs as well as multiple outputs.

1.4.4 General Model

- The basic distinction of Mealy machine is that the outputs of the outside world are function of two state of variables as under :
    1. The present input condition.
    2. The present state of the machine.

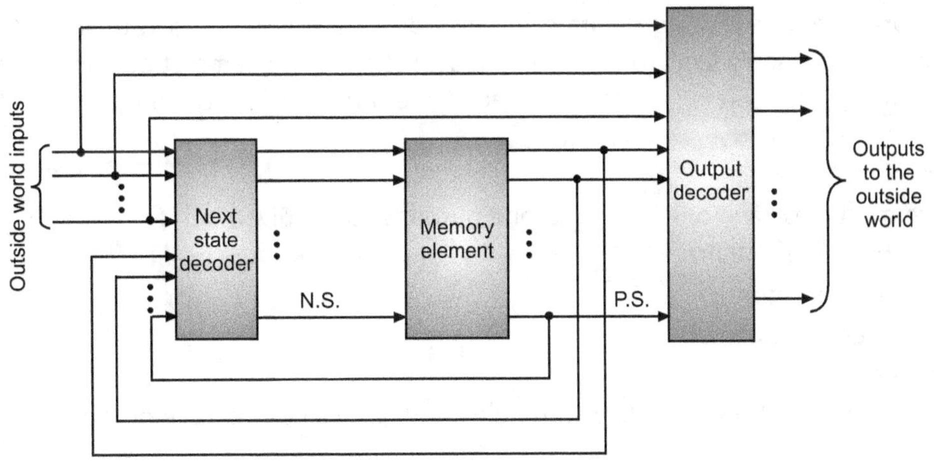

Fig. 1.3 : Block Diagram for General Model of a Mealy Machine

- Mealy machine model is derived from the general basic model shown in Fig. 1.3 by a process of generation.
- Mealy machine is a class A machine, which is named after G. H. Mealy who is one of the pioneer of sequential circuits.
- A block diagram of Mealy machine model is shown in Fig. 1.3.

1.4.5 Applications

1. Number classification
2. Watch with timer
3. Vending machine
4. Traffic light
5. Bar code scanner
6. Gas pumps
7. Describing engine
8. Modern CPUs
9. Cell phones
10. Digital clocks
11. Computers

1.5 MOORE MACHINE

1.5.1 Introduction

- In the **theory of computation**, a Moore machine is a **finite-state machine,** whose output values are determined only by its current state. This is in contrast to a **Mealy machine**, whose (Mealy) output values are determined both by its current state and by the values of its inputs.
- The Moore machine is named after Edward F. Moore, who presented the concept in 1956.

1.5.2 Definition

- **Moore machine is defined as a finite state machine in which the next state is decided by the current state and current input symbol.**
- The output symbol at a given time depends only on the present state of the machine.
- A Moore machine can be defined as a 6-tuple $(S, S_0, \Sigma, A, T, G)$ consisting of the following:
  - A finite set of states S
  - A start state (also called initial state) S_0 which is an element of S
  - A finite set called the input alphabet Σ
  - A finite set called the output alphabet A
  - A transition function $T : S \times \Sigma \rightarrow S$ mapping a state and the input alphabet to the next state
  - An output function $G : S \rightarrow A$ mapping each state to the output alphabet

1.5.3 State Diagram

- The state transition table is a table showing relation between an input and a state.
- The state diagram for a Moore machine or Moore diagram is a diagram that associates an output value with each state. It is an output procedure.

- When it is represented as a state diagram, each node (state) is labelled with an output value. Every Moore machine M is equivalent to the Mealy machine with the same states and transitions and the output function G(s, σ) → G_M (s), which takes each state-input pair (s, σ) and yields G_M (s), where G_M is M's output function.

1. Simple Example:

- Simple Moore machines have one input and one output such as given below:
    - (i) Edge detector using EX-OR.
    - (ii) Binary adding machine.
    - (iii) Clocked sequential systems.

2. Complex Examples:

- More complex Moore machines can have multiple inputs as well as multiple outputs.

1.5.4 General Model

- Moore machine is the class B and class C machines, and named after E. F. Moore, who is another pioneer of sequential circuits.
- Moore machine model is derived from the general basic model shown in Fig. 1.4 by a process of generation.
- The basic distinction of a Moore machine is that its output is strictly function of the state of the machine.

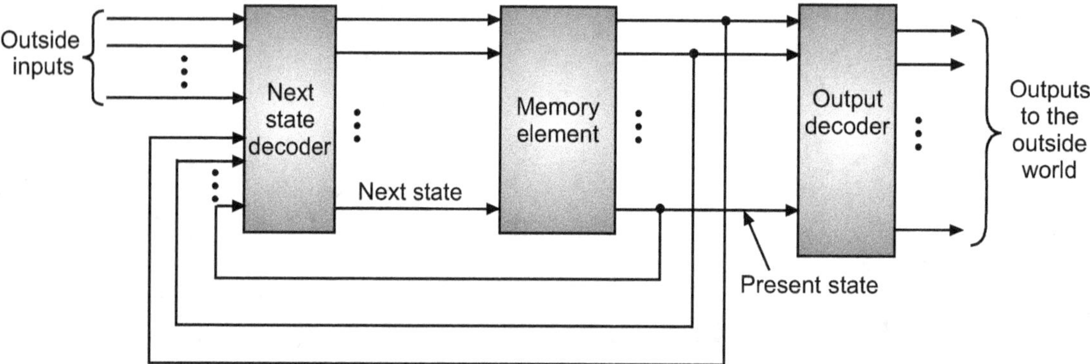

Fig. 1.4 : Block diagram for a general model of a Moore machine

1.5.5 Applications

1. Number classification.
2. Watch with timer.
3. Traffic light.
4. Bar code scanner.
5. Gas pumps.

1.6 COMPARISON OF MEALY AND MOORE MACHINE

Table 1.1

Mealy Machine	Moore Machine
1. Output depends on present state as well as present input.	1. Output depends only upon present state.
2. If input changes, output also changes.	2. If input changes, output does not change.
3. Less number of states are required.	3. More number of states are required.
4. There is less hardware requirement.	4. There is more hardware requirement.
5. They react faster to inputs.	5. They react slower to inputs (one clock cycle later).
6. Asynchronous output generation.	6. Synchronous output and state generation.
7. Output is placed on transitions.	7. Output is placed on states.
8. It is difficult to design.	8. Easy to design.

1.7 IMPLEMENTATION OF SEQUENTIAL CIRCUITS

1.7.1 Description of State

- Sequential circuits have the capability to retain the effect of all part inputs on present and future output.
- The memory elements retain the complete sequence of inputs at a time t_o and able to determine the output at any time $t > t_p$.
- For simplification, the values of these input sequences and be grouped into a finite number of classes in such a way that all time functions having the same effect on the output at time $t \geq t_o$ are included in the same class.
- So the determination of output does not require the whole input sequence, because knowing the class of the function is enough to determine the output.
- The class is kept in an auxiliary variable S called state. Since the number of classes are finite, such systems are also called finite state machines.
- The state description of a sequential circuit use three variables, i.e. the input, the state and the output.
- There are two state transition functions as under :
  1. The state transition function, which produces the next state (at time $t + 1$) as a function of the present input $X(t)$ and present state $S(t)$.
  2. The output function which produces the present output $Z(t)$ as a function of the present and next state.
- Here, $S(t)$ is known as the **present state** and $S(t + 1)$ is known as the **next state**.
- The transitions and output functions are stored in a **state table**. The state table consists of first listing of all possible binary combination of present state and inputs.
- The next state values are then determined from the logic diagram or from the state sequences.
- For each present state, there are two possible next states and outputs, depending on values of the input.
- The information available in a state table can be represented **graphically** in a state diagram.
- An combinational circuit can be described in a tabular format called **truth table**, similarly the behaviour of sequential circuits can be described in a tabular format called **state diagram.**

1.7.2 State Diagram

Definition :

- *The graphical representation of state to state transitions of a sequential circuit is known as the state diagram.*
- *The state diagram is an array of '**bubbles**' connected by directed line sequence with arrowheads.*
- Each bubble represents a state of the machine and the line segments are graphical indication of state diagram.
- The five important and descriptive pieces of information for each state are given and are labelled as under :
  1. Some state identifying symbol or code.
  2. The previous state of the state-of-interest.
  3. The input condition leading to the state of interest.
  4. The output specification for the state of interest.
  5. The next states and next state branching conditions for the state-of-interest.

1.7.3 Examples

- The example of a state diagram is shown in Fig. 1.5 which illustrates the state of interest and five important and required pieces of information related to each state of the machine.

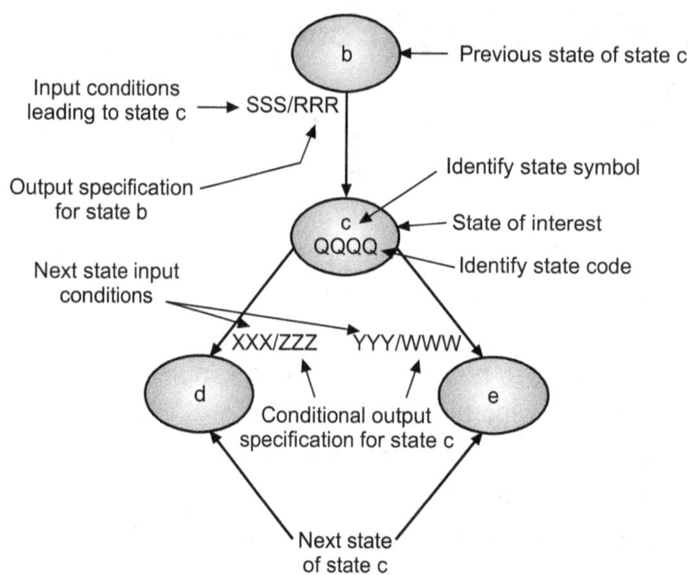

Fig. 1.5 : A segment of a state diagram

1.7.4 Implementation of Mealy Machine

1. State diagram :

- The state diagram of Mealy machine is shown in Fig. 1.6.

- The output signals are dependent on both present state and all input signals. It means the output signals change immediately if the input signals change or if the state is changed.

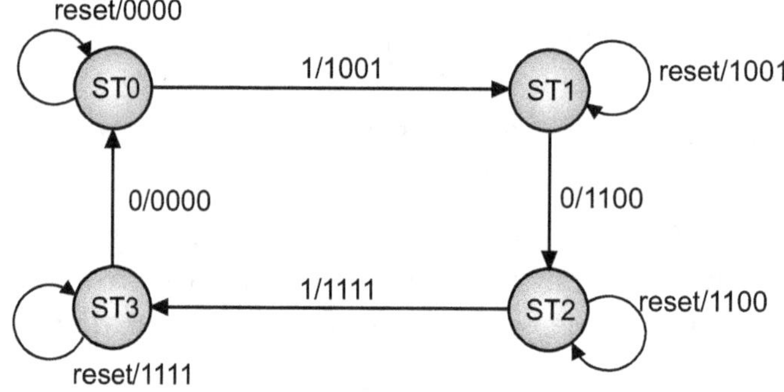

Fig. 1.6 : Mealy machine's state diagram

2. Block diagram :

- The block diagram of Mealy machine is shown in Fig. 1.7.

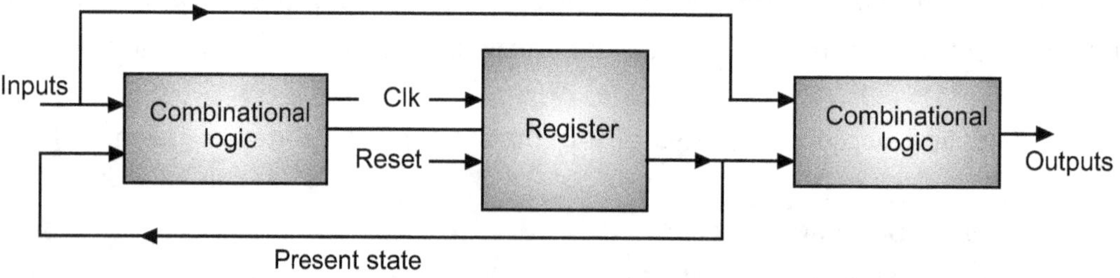

Fig. 1.7 : Mealy machine's block diagram

1.7.5 Implementation of Moore Machine

1. State diagram :

- It is noted that the output signals are only dependent on the present state for the Moore machine.

- The output signals are usually drawn inside the state bubbles. The state diagram of Moore machine is shown in Fig. 1.8.

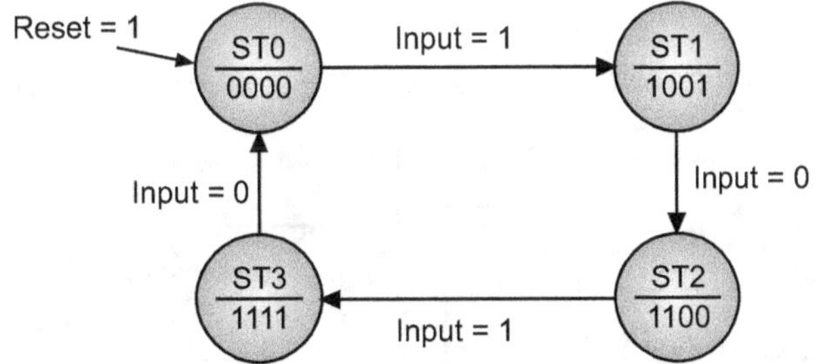

Fig. 1.8 : Moore machine's state diagram

2. Block diagram :

- The block diagram of Moore machine is shown in Fig. 1.9.

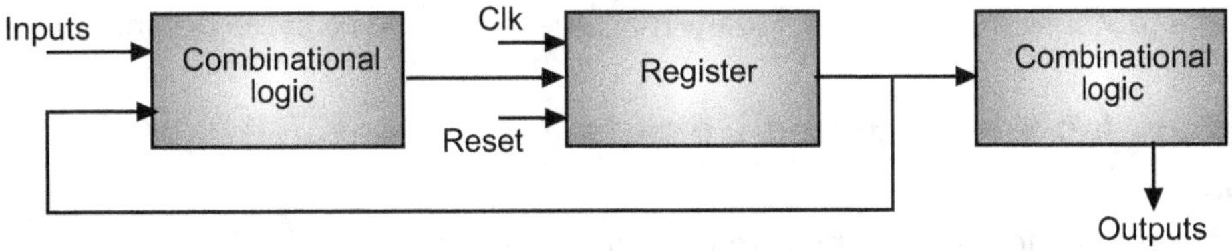

Fig. 1.9 : Moore machine's block diagram

1.7.6 Disadvantage of Mealy and Moore Machines

- The output signals for Mealy and Moore machines are from combinational logic. The state machine can have spike free outputs at a certain temperature or supply voltage.

- But as temperature or supply voltage changes, it generates spikes in the output signals. Normally such spikes are not important.

- In synchronous design of combinational logic, it is only at the active clock edge that the data signal must be stable.

1.7.7 Methods for Spike Free Outputs

- The following methods can be used for spike free output signals of Mealy and Moore machines :

   1. Output-state machine.

   2. Mealy machine with clocked outputs.

   3. Moore machine with clocked outputs.

1.8 MEALY AND MOORE MACHINES WITH CLOCKED OUTPUTS

1.8.1 Mealy Machine with Clocked Outputs   (W-16)

1. State diagram :

- Fig. 1.10 shows the state diagram of Mealy machine with clocked outputs.

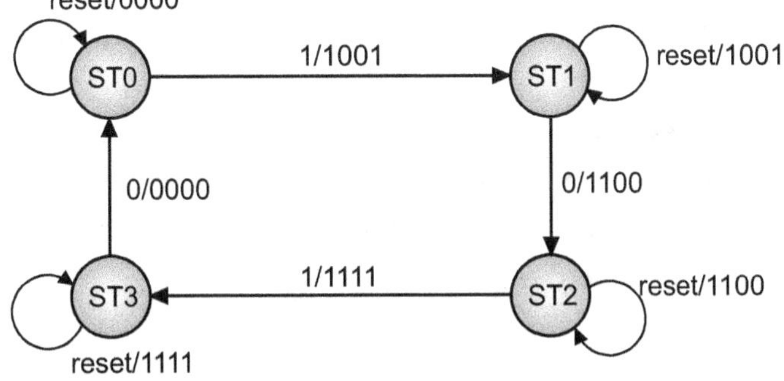

Fig. 1.10 : State diagram for a Mealy machine with clocked outputs

2. Block diagram :

- A block diagram of Mealy machine with clocked output is shown in Fig. 1.11.

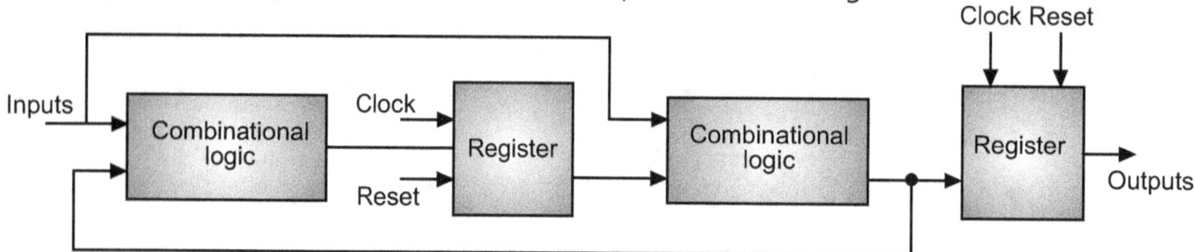

Fig. 1.11 : Block diagram for a Mealy machine with clocked outputs

- With clocked outputs, all the outputs are synchronized with an extra D-type flip-flop to obtain spike free outputs.

- In this also, the outputs in a clocked Mealy machine are a clock cycle behind the ordinary Mealy's machine.

1.8.2 Moore Machine with Clocked Outputs   (W-16)

1. State diagram :

- The state diagram of a Moore machine with clocked outputs is shown in Fig. 1.12.

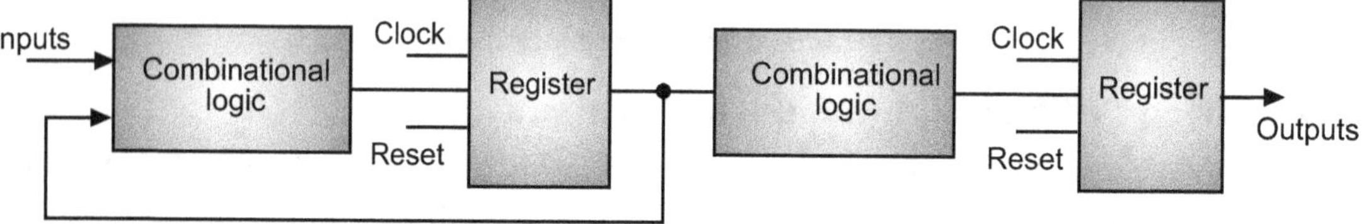

Fig. 1.12 : Block diagram of a Moore machine with clocked outputs

2. Block diagram :

- A block diagram of a Moore machine with clocked outputs is shown in Fig. 1.13.

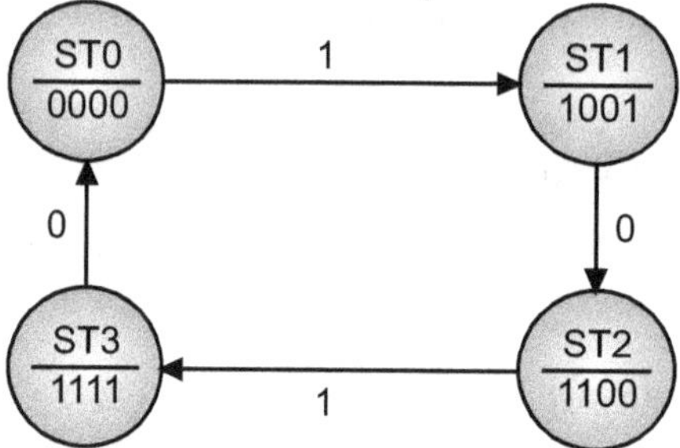

Fig. 1.13 : State diagram of a Moore machine with clocked outputs

- With clocked outputs all the outputs are synchronized with extra D-type flip-flop to obtain spike free outputs. It means that the outputs are a clock cycle behind the ordinary Moore machine's output.

1.9 SEQUENCE DETECTOR

1.9.1 Definition

- A **sequence detector** is a sequential state machine. In a Moore machine, output depends only on the present state and not dependent on the input (x). Hence in the diagram, the output is written with the states.

- **A sequence detector is a sequential state machine, which takes an input string of bits and generates an output1, whenever the target sequence has been detected.**

- In a Mealy machine, output depends on the present state and the external input x. Hence in the diagram, the output is written outside the states.

1.9.2 Types

1. Overlapping
2. Non-overlapping

- In an overlapping sequence detector the last bit of one sequence becomes the first bit of next sequence. However, in non-overlapping sequence detector the last bit of one sequence does not become the first bit of next sequence

1.9.3 Design of Mealy Sequence Detector

- The steps to design non-overlapping 101 Mealy sequence detector are:

Step 1: Develop the state diagram:

The state diagram of a Mealy machine for a 101 sequence detector is as shown in Fig. 1.14.

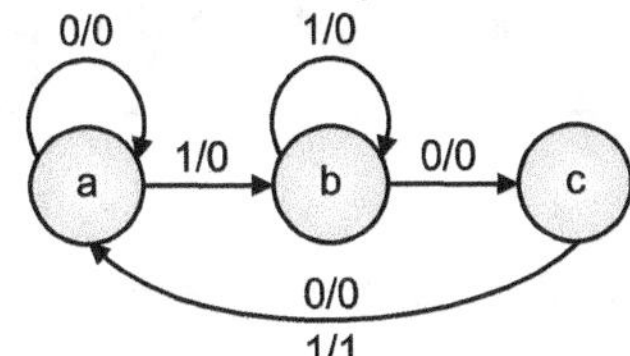

Fig. 1.14 : State diagram of a 101 sequence detector

Step 2: Code Assignment:

Rule 1: States having the same next states for a given input condition should have adjacent assignments.

Rule 2: States that are the next states to a single state must be given adjacent assignments.

Rule 1 is given preference over Rule 2.

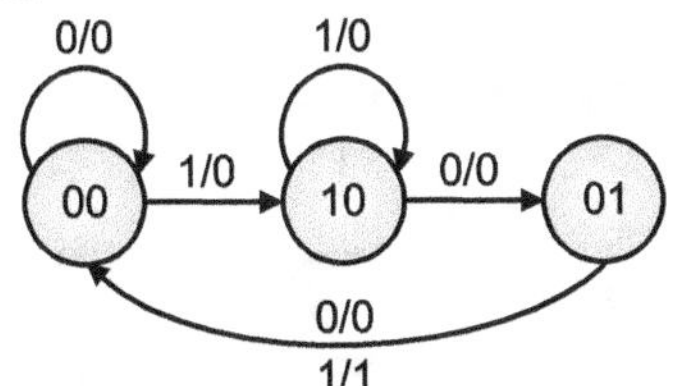

Fig. 1.15 : State diagram after code assignment

The state diagram after the code assignment is as shown in Fig. 1.15.

Step 3: Make Present State/Next State table:

We will use D-flip flops for design purpose.

Table 1.2

Present states		Input	Next states		Flip Flop Excitations		Output
X	Y		X'	Y'	D_x	D_y	
0	0	0	0	0	0	0	
0	0	1	1	0	1	0	
0	1	0	0	0	0	0	
0	1	1	0	0	0	0	
1	0	0	0	1	0	1	
1	0	1	1	0	1	0	
1	1	0	x	x	x	x	
1	1	1	x	x	x	x	

Step 4: Draw K-maps for D_x, D_y and output (Z):

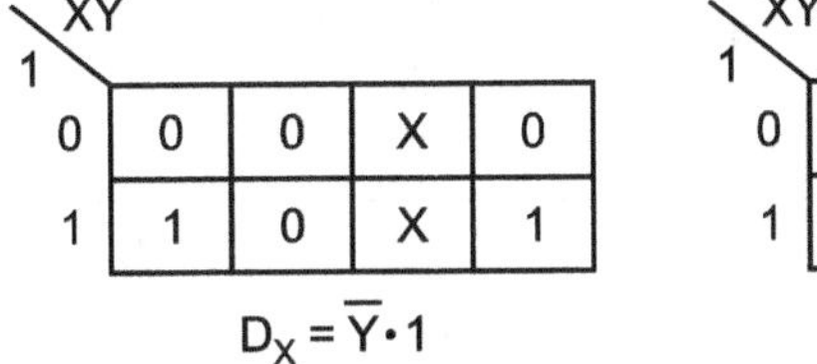
$$D_X = \overline{Y} \cdot 1$$

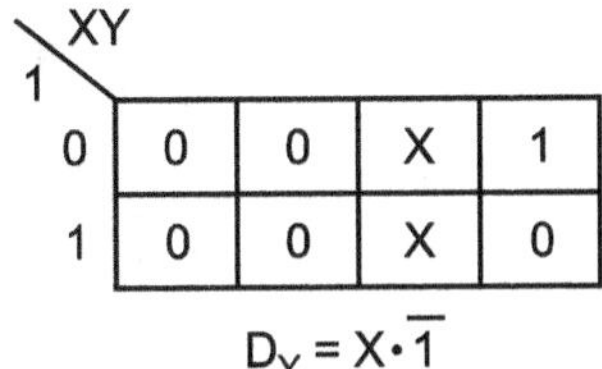
$$D_Y = X \cdot \overline{1}$$

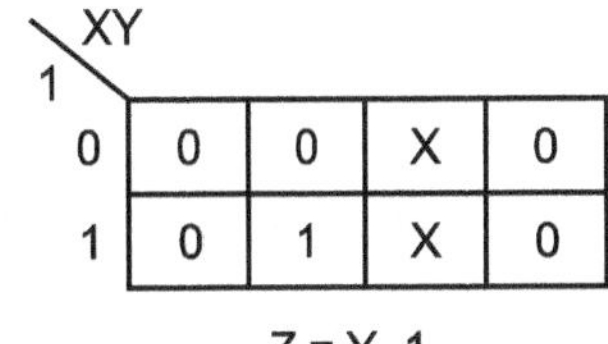
$$Z = Y \cdot 1$$

Fig. 1.16 : K-maps for D_x, D_y and Z

Step 5: Finally implement the circuit:

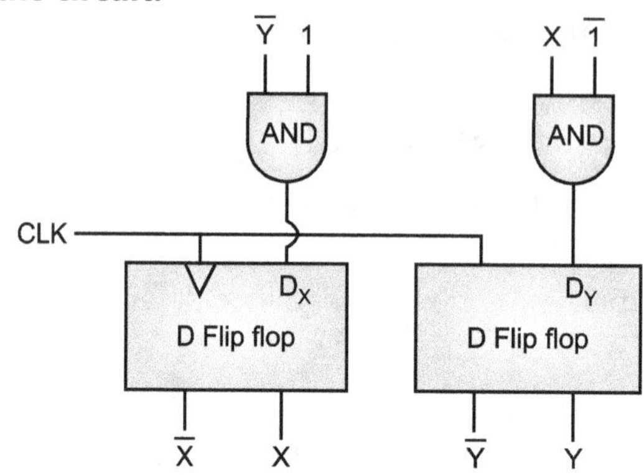

Fig. 1.17 : Mealy 101 non-overlapping sequence detector

This is the final circuit for a Mealy 101 non-overlapping sequence detector.

1.9.4 Design of Moore Sequence Detector

- The state diagram of a Moore machine for a 101 detector is as shown in Fig. 1.18.

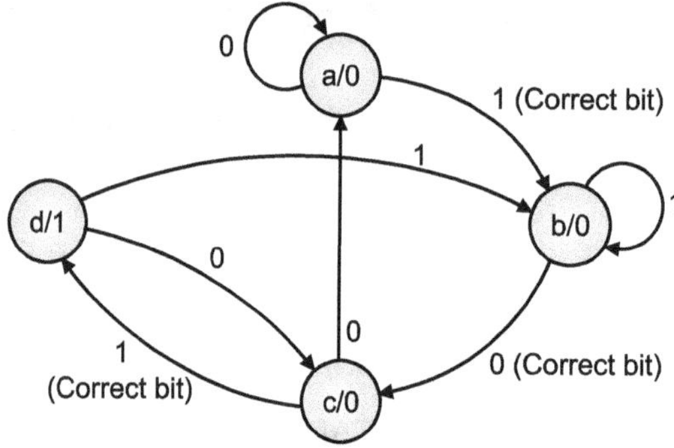

Fig. 1.18 : State diagram for 101 Moore sequence detector

- The state table for Fig. 1.18 is shown in Table 1.3.

Table 1.3

Present state	Next state		Output	
	x = 0	**x = 1**	**x = 0**	**x = 1**
a	a	b	0	0
b	c	b	0	0
c	a	d	0	1
d	c	b	0	0

- Four states will require two flip flops. Consider two D flip flops. Their excitation table is shown below in Table 1.4.

Table 1.4 : Excitation Table

Input	Present state		Next state		F/F inputs		Output
x	**B**	**A**	**B+1**	**A + 1**	**D_x**	**D_y**	**Y**
0	0	0 (a)	0	0 (a)	0	0	0
0	0	1 (b)	1	0 (c)	1	0	0
0	1	0 (c)	0	0 (a)	0	0	0
0	1	1 (d)	1	0 (c)	1	0	0
1	0	0 (a)	0	1 (b)	0	1	0
1	0	1 (b)	0	1 (b)	0	1	0
1	1	0 (c)	1	1 (d)	1	1	0
1	1	1 (d)	0	1 (b)	0	1	1

- K-maps to determine inputs to D flip flop.

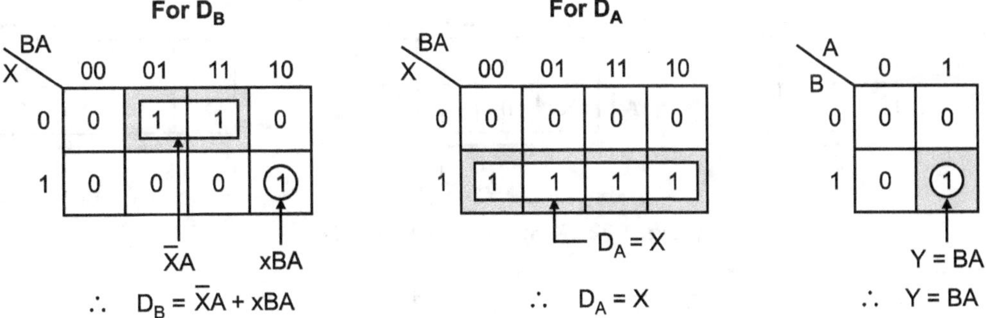

$$\therefore \quad D_B = \overline{X}A + xBA \qquad \therefore \quad D_A = X \qquad \therefore \quad Y = BA$$

Fig. 1.19 : K-maps

- Circuit diagram for the sequence detector:

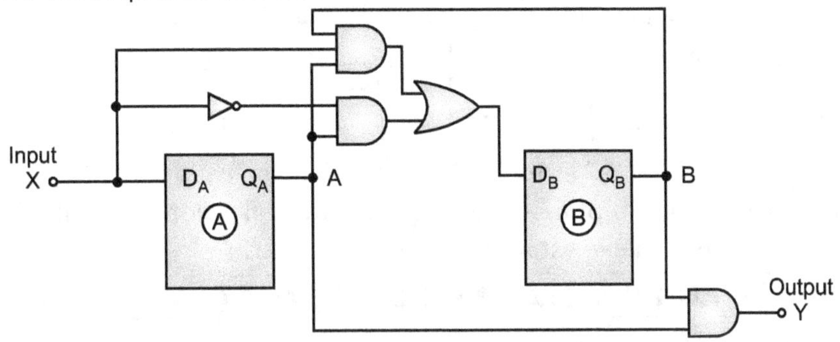

Fig. 1.20 : Moore non-overlapping 101 sequence detector

1.10 FSM COUNTERS

1.10.1 Introduction

- Synchronous counters are potentially attractive for implementing finite state machines. In a single package they provide a state register with mechanisms for advancing the state (CNT), resetting it to zero (CLR), and "jumping" to a new state (LD). Rather than encode the next-state function as actual state bits, you can implement it by asserting the counter control signals CNT, LD, and CLR under the right conditions.

- For correct operation in state sequencing, the counter must be implemented with synchronous LD and CLR signals, such as the TTL 74163 binary up-counter. As pointed out in previous section, asynchronous control signals lead to invalid behavior of the state machine.

1.10.2 Design of 5 State Counter

- Counter repeats in sequence of 5 states as 000, 010, 011, 101, 110, 000.

Step 1: State diagram   **Step 2: State transition table**

Table 1.5

Present state			Next state		
C	B	A	C+	B+	A+
0	0	0	0	1	0
0	0	1	X	X	X
0	1	0	0	1	1
0	1	1	1	0	1
1	0	0	X	X	X
1	0	1	1	1	0
1	1	0	0	0	0
1	1	1	X	X	X

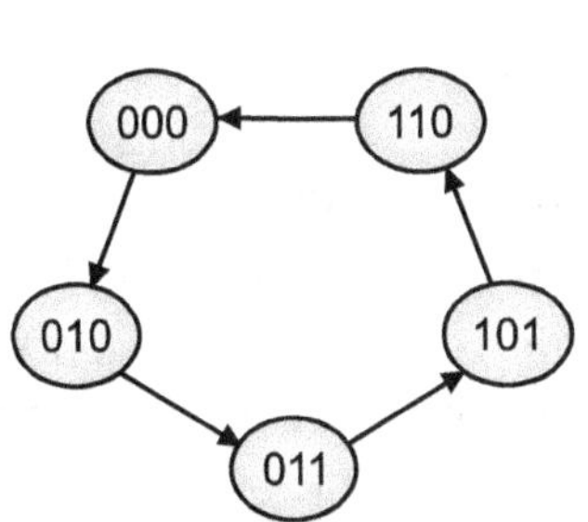

Fig. 1.21 : State diagram

Step 3: Encode the next functions

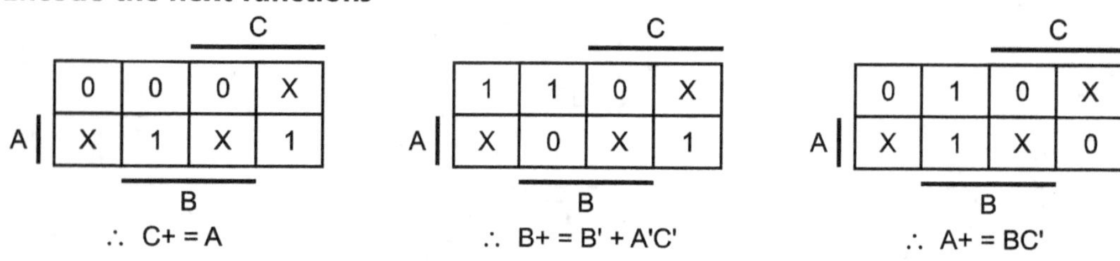

$$\therefore\ C+ = A \qquad\qquad \therefore\ B+ = B' + A'C' \qquad\qquad \therefore\ A+ = BC'$$

Fig. 1.22 : K-maps for A+, B+, C+

Step 4: Implement the design

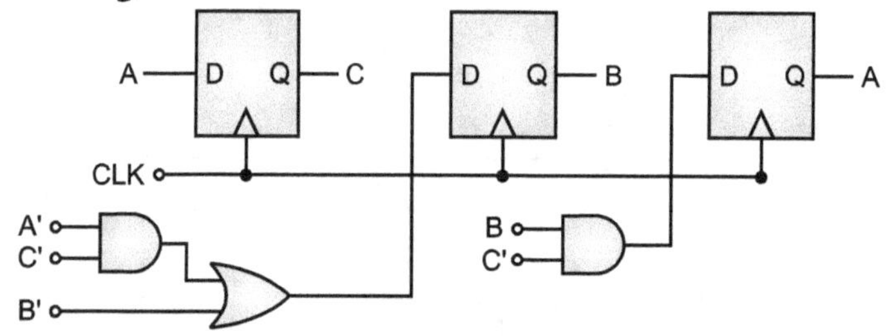

Fig. 1.23 : 5-State Counter with D flip flops

- Fig. 1.23 shows the circuit diagram of a 5-state counter with D flip flops. A D flip flop also produces Q' so A', B', and C' would all be available without any extra inverters.

1.11 APPLICATION SPECIFIC INTEGRATED CIRCUIT (ASIC)

1.11.1 Introduction

- The initial ASICs used gate array technology. An early successful commercial application was the gate array circuiting found in the low end 8-bit personal computers, introduced in 1981 and 1982.

- Customization occurred by varying the metal interconnect mask. Gate arrays had complexities of up to a few thousand gates, this is now called mid-scale integration (MSI).

- The maximum complexity and hence functionality possible in an ASIC has grown from 5,000 logic gates to over 100 million. Modern ASICs often include entire microprocessors, memory blocks including ROM, RAM, EEPROM, flash memory and other large building blocks.

- Such an ASIC is often termed as SoC (system-on-chip). Designers of digital ASICs often use a *hardware description language* (HDL), such as *Verilog* or *VHDL*, to describe the functionality of ASICs.

1.11.2 Definition

- **An ASIC is an integrated circuit (IC) chip customized for a particular use, rather than intended for general-purpose use.**

- For example, a chip designed to run in a digital voice recorder or a high-efficiency bitcoin miner is an ASIC.

- **An ASIC is a microchip designed for a special application, such as a particular kind of transmission protocol or a hand-held computer.**

1.11.3 Concept

- To design a digital circuit, a designer can select an appropriate IC for the circuit due to its advantages such as low development cost, fast turn around of designs and relatively easy to test the circuits.

- But the disadvantages of these design methods are large area (size) of PCB requirements, large power requirements, less security and additional cost, space, power requirements etc. required to modify the design.

- To overcome the disadvantages of design using fixed function ICs, Application Specific Integrated Circuits (ASICs) have been developed.

- The ASICs are designed by the users to meet the specific requirements of a circuit and are produced by an IC manufacturer as per the specifications supported by the user.

- *ASIC stands for Application Specific Integrated Circuit. It implements digital circuit functions addressing to system requirements. It is useful for high volume production.*

1.11.4 Applications

1. Auto emission control
2. Environment monitoring
3. Personal digital assistants
4. Computer's graphics
5. Modem
6. Encoding/decoding stand done USB interface chip

1.11.5 Classification

- ASICs can be classified into two basic categories as under :
  1. Full custom ASICs.
  2. Semi-custom ASICs.
     - (a) Standard cells based ASICs.
     - (b) Gate array based ASICs.
     - (c) Moved signal and analogue ASICs.
     - (d) Programmable logic devices.

1.11.6 Full Custom ASIC

- *Full custom ASIC is an IC designed by using basic logic gates, circuits or layout specifically for a particular design.*
- Here all logic-cells are customized by the designer. The designer abandons the approach of using pretested and precharacterized cells for his design. Every mask is designed by the customer.
- The CMOS technology is used for full custom because it is easier to design analogues well as digital circuits on the same chip with better performance.
- These are designed and processed like standard products with all set of masks for all fabrication layers e.g. mega processor like pentium, memory etc.

1.11.7 Semi-custom ASIC

1. Definition :

- *Semi-custom ASICs are either partly or fully prefabricated or they are structured like standard and custom ICs.*
- Few mask patterns are predesigned.

2. Classification :

- ASICs can be classified into four categories as under :
  - (a) Standard cell-based ASICs.
  - (b) Gate array based ASICs.
  - (c) Mixed signal and analogue ASICs.
  - (d) Programmable logic devices.

1.11.8 Standard Cell Based ASIC

1. Concept :

- *The predesigned logic cells such as gates, multipliers, flip-flops etc. are known as standard cells.*
- The standard cells are used in standard cell-based ASICs.
- Standard cell areas in a cell-based IC (CBIC) are built of rows of standard cells like a wall built of bricks.
- Standard cell areas may be used in combination with larger predesigned cells, perhaps microcontrollers or even micro-processors known as mega cells.
- All mask layers are customized this means that the standard cells can be placed any where on the silicon.

2. Standard cell structure :

- Fig. 1.24 shows the layout of a typical standard cell.
- Standard cell is fabricated with full set of masks. The design support includes a library of macro cells. The macro contain the patterns for all the masks and the design is completed by the placement and routing which determines the location of the macros and their interconnections.

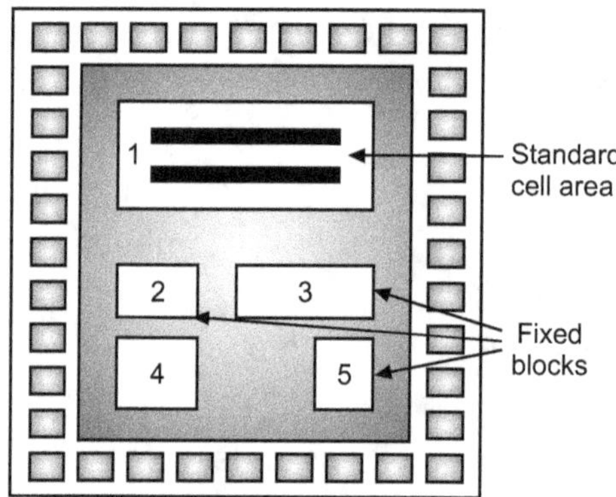

Fig. 1.24 : Structure of standard cells

- The power and ground lines run parallel to the upper and lower boundaries of the cells, hence neighbouring cells share a common power and ground bus.
- The input and output pins are located on the upper and lower boundaries of the cells.

3.   Features :

- The features of standard cell-based ASICs are as under :
    (a) All mask layers are customized.
    (b) Custom blocks can be embedded.
    (c) Manufacturing lead-time is about few weeks.

1.11.9 Gate Array Based ASIC

1.   Concept :

- In gate array-based ASICs, pattern of transactions are predefined on the silicon.
- The predefined pattern of transistors on a gate array is the base array and the smallest element that is replicated to make the base array is the base cell or primitive cell.
- The designer chooses from a library of logic cell. The logic cells in a gate array library are called macros.
- *A gate array ASIC is an IC chip on which gates are placed in matrix form without connections among the gates.*

2.   Types of gate arrays :

- The types of gate arrays are as under :
    1.   Channeled gate arrays.
    2.   Channelless gate array or sea of gate array.

1.11.10 Channel Gate Array ASIC

1.   Features :

- The features of channeled gate array-based ASICs are :
    1.   Only the interconnect is customized.
    2.   The interconnect uses predefined spaces between rows of base cells.
    3.   Manufacturing time is about few days.

2.   Structure :

- An structure of channeled gate array based ASIC is shown in Fig. 1.25.

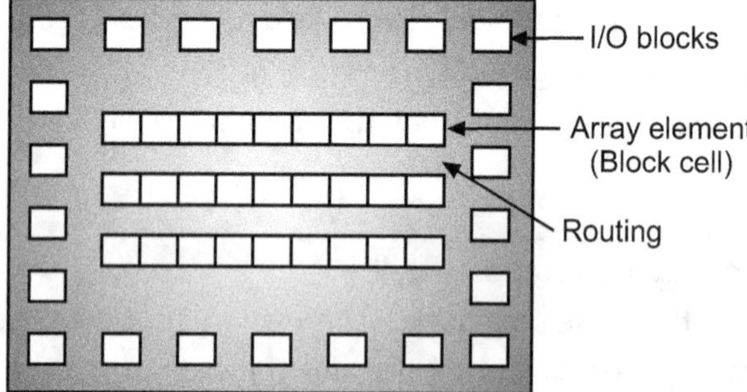

Fig. 1.25 : Structure of channeled gate array ASIC

- The rows of basic cells with spaces are occupied on the silicon core.
- *The space between the array element is called routing channel.*
- The routing channel is kept free for the interconnection among the basic cells.
- The core array is supplemented by the peripheral input/output blocks which can be configured into input, output or bidirectional buffers.
- An array element consists of two PMOS and two NMOS. The chip function is implemented by the metallization, which interconnects selected array element.

1.11.11 Channelless Gate Array ASIC

- Architecture of channelless gate array or sea of gate (SOG) ASIC is shown in Fig. 1.26.

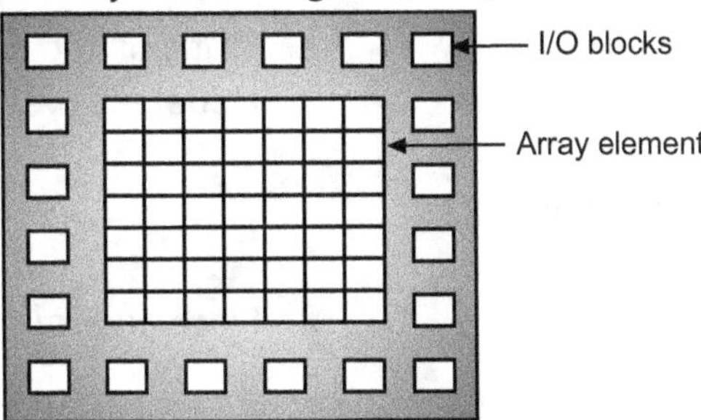

Fig. 1.26 : Structure of channelless gate array ASIC

- It occupies the entire core of the chip. All interconnects are now passed over array element. The number of array elements are increased.
- There are no predefined areas for routing between cells. So, SOG architecture allows larger number of array elements to be utilized than a channeled array for the same size of die, i.e. the logic density is higher.

1.11.12 Mixed Signal and Analog ASIC

- The mixed mode and analog ASICs contain analog and digital circuits.
- The analog section covers functions like operational amplifiers, comparators, voltage references and data converters.
- Standard cell formed is used to construct analog circuits.
- Analog and digital macros are sided by in a mixed mode ASIC. They are formed by interconnecting a linear array, which contains an assembly of bipolar transistors, resistors and MOS capacitors.
- The vendor supplies the data for interconnections.

1.12 ASIC DESIGN

1.12.1 Introduction

- Today, ASIC design flow is a mature process with many individual steps. ASIC design flow process is the backbone of every ASIC design project. To ensure design success, one must have: a silicon-proven ASIC design flow, a good understanding of the ASIC specifications and requirements, and an absolute domination over the required EDA tools (and their inputs and outputs).

1.12.2 Why to Adopt/Need

- To ensure successful ASIC design, engineers must follow a proven ASIC design flow, which is based on a good understanding of ASIC specifications, requirements, low power design, and performance, with a focus on meeting the goal of the right time to market. Every stage of the ASIC design cycle has EDA tools that can help to implement ASIC design with ease.

1.12.3 Design Flow Steps

1. Logic synthesis
2. Floor planning
3. Synthesis
4. Block level layout
5. ASIC level layout

Step 1: Logic Synthesis Steps
1. RTL conversion into netlist,
2. Design partitioning into physical blocks,
3. Timing margin and timing constraints,
4. RTL and gate level netlist verification,
5. Static timing analysis.

Step 2: Floorplanning Steps
1. Hierarchical ASIC blocks placement
2. Power and clock planning

Step 3: Synthesis Steps
1. Timing constrains and optimization
2. Static timing analysis
3. Update placement
4. Update power and clock planning.

Step 4: Block Level Layout Step
1. Complete placement and routing of blocks.

Step 5: ASIC Level Layout Step
1. ASIC integration of all blocks,
2. Place and route,
3. GDSCI creation

1.12.4 Working of Principle

- In order to fulfill the futuristic demands of chip design, changes are required in design tools, methodologies, and software/hardware capabilities. For these changes, ASIC design flow adopted by engineers for efficient structured ASIC chip architecture and focus on its design functionalities
- ASIC design flow is a mature and silicon-proven IC design process, which includes various steps like design conceptualization, chip optimization, logical/physical implementation, and design validation and verification.

Step 1: Chip Specification

- This is the stage at which the engineer defines features, microarchitecture, functionalities (hardware/software interface), specifications (Time, Area, Power, Speed) with design guidelines of ASIC. Two different teams are involved at this juncture:
  o Design team: Generates RTL code.
  o Verification team: Generates test bench.

Step 2: Chip Partitioning

- This is the stage wherein the engineer follows the ASIC design layout requirement and specification to create its structure using EDA tools and proven methodologies. This design structure is going to be verified with the help of HLL programming languages like C++ or System C.
- After understanding the design specifications, the engineers partition the entire ASIC into multiple functional blocks (hierarchical modules), while keeping in mind ASIC's best performance, technical feasibility, and resource allocation in terms of area, power, cost, and time. Once all the functional blocks are implemented in the architectural document, the engineers need to brainstorm ASIC design partitioning by reusing IPs from previous projects and procuring them from other parties.

Step 3: Design Entry/Functional Verification

- Functional verification confirms the functionality and logical behavior of the circuit by simulation on a design entry level. This is the stage where the design team and verification team come into the cycle where they generate RTL code using test-benches. This is known as **behavioral simulation**.

Step 4: RTL Block Synthesis/RTL Function

- Once the RTL code and testbench are generated, the RTL team works on RTL description – they translate the RTL code into a gate-level netlist using a logical synthesis tool that meets required timing constraints. Thereafter, a synthesized database of the ASIC design is created in the system. When timing constraints are met with the logic synthesis, the design proceeds to the design for testability (DFT) techniques.

Step 5: Design for Test (DFT) Insertion

- With the ongoing trend of lower technology nodes, there is an increase in system-on-chip variations like size, threshold voltage, and wire resistance. Due to these factors, new models and techniques are introduced to high-quality testing.
- ASIC design is complex enough at different stages of the design cycle. Telling the customers that the chips have a fault when you are already at the production stage is embarrassing and disruptive. It is a situation that no engineering team wants to be in. In order to overcome this situation, design for test is introduced with a list of techniques:
  1. Scan path insertion
  2. Memory BIST (built-in Self-Test)
  3. ATPG (automatic test pattern generation)

Step 6: Floor Planning (Blueprint Your Chip)

- After DFT, the physical implementation process is to be followed. In physical design, the first step in RTL-to-GDSII design is floorplanning. It is the process of placing blocks in the chip. It includes block placement, design portioning, pin placement, and power optimization.
- Floorplan determines the size of the chip, places the gates, and connects them with wires. While connecting, engineers take care of wire length and functionality, which will ensure signals will not interfere with nearby elements. In the end, simulate the final floor plan with the post-layout verification process.
- A good floorplanning exercise should come across and take care of the below points; otherwise, the life of IC and its cost will blow out:
  1. Minimize the total chip area
  2. Make routing phase easy (routable)
  3. Improve signal delays

Step 7: Clock Tree Synthesis

- Clock tree synthesis is a process of building the clock tree and meeting the defined timing, area, and power requirements. It helps in providing the clock connection to the clock pin of a sequential element in the required time and area, with low power consumption.
- In order to avoid high power consumption, increase in delays, and a huge number of transitions, certain structures can be used for optimizing CTS structure such as Mesh Structure, H-Tree Structure, X-Tree Structure, Fishbone Structure, and Hybrid structure.
- With the help of these structures, each flop in the clock tree gets the clock connection. During the optimization, tools insert the buffer to build the CTS structure. Different clock structures will build the clock tree with a minimum buffer insertion and lower power consumption of chips.

Step 8: Place and Route

- As we are moving towards a lower technology node, engineers face complex design challenges with the need for implanting millions of gates in a small area. In order to make this ASIC design routable, placement density range needs to be followed for better QoR. Placement density analysis is an important parameter to get better outcomes with less number of iterations.
  1. **Global Routing**: Calculates estimated values for each net by the delays of fan-out of wire. Global routing is mainly divided into line routing and maze routing.
  2. **Detailed Routing**: In detailed routing, the actual delays of wire is calculated by various optimization methods like timing optimization, clock tree synthesis, etc.

Step 9: Final Verification (Physical Verification and Timing)

- After routing, the ASIC design layout undergoes three steps of physical verification, known as sign-off checks. This stage helps to check whether the layout working the way it was designed to. The following checks are followed to avoid any errors just before the tape-out:

- Layout versus schematic (LVS) is a process of checking that the geometry/layout matches the schematic/netlist.

- Design rule checks (DRC) is the process of checking that the geometry in the GDS file follows the rules given by the foundry.

- Logical equivalence checks (LVC) is the process of equivalence check between pre and post design layout.

Step 10: GDS II – Graphical Data Stream Information Interchange

- In the last stage of the tapeout, the engineer performs wafer processing, packaging, testing, verification and delivery to the physical IC. GDSII is the file produced and used by the semiconductor foundries to fabricate the silicon and handled to client.

- EInfochips has contributed to over 500 product designs for top global companies, with more than 40 million deployed around the world. As a leading ASIC design and verification service provider, eInfochips has brought together IP cores, verification IP and design and verification expertise.

1.13 PROGRAMMABLE LOGIC DEVICES (PLDs)

1.13.1 Introduction

- A logic device is an electronic component, which performs a definite function which is decided at the time of manufacture and will never change. For example, a NOR gate always inverts the logic level of the input signal and does/can-do-nothing else.

- On the other hand, **Programmable Logic Devices** (PLDs) are the components, which do not have a specific function associated with them. These can be configured to perform a certain function by the user, on a need basis and can further be changed to perform some other function at the later point of time, i.e. these are re-configurable. However, the amount of flexibility offered depends on their type.

- In 1969, **Motorola** offered the XC157, a mask-programmed gate array with 12 gates and 30 uncommitted input/output pins.

- In 1970, **Texas Instruments** developed a mask-programmable IC based on the IBM read-only associative memory or ROAM.

- In 1971, General Electric Company (GE) was developing a programmable logic device based on the new **Programmable Read-Only Memory** (PROM) technology.

1.13.2 Definition

- *A programmable logic device (PLD) is an electronic component used to build reconfigurable digital circuits.*

- Unlike integrated circuits (IC) which consist of logic gates and have a fixed function, a PLD has an undefined function at the time of manufacture.

- PLDs are configured IC programmed to create a part customized to a specific application and so they also belong to the family of ASIC.

- **Programmable logic devices (PLDs) are the ICs, which contain an array of AND gates and another array of OR gates.**

- The process of entering the information into these devices is known as **programming**.

- **Programmable Logic Devices (PLDs) are the components which do not have a specific function associated with them. These can be configured to perform a certain function by the user.**

1.13.3 Need

- The technological growth in digital design has increased complexity of the system to a large extent. Also size, power and speed are the major factors in the design aspect. At the same time, PLDs serve the many advantages such as low cost, design security, greater reliability, flexibility, automation etc. Therefore, there is a need of PLDs in digital design.

1.13.4 Features

1. No customized mask layers in logic cells.
2. Past design turn around time.
3. A single large block of programmable interconnect.

1.13.5 Advantages

1. Design flexibility,	2. Improved reliability
3. Lower power consumption,	4. Reduced complexity
5. Field programmable,	6. Erasable,
7. Reprogrammable	

- The advantages of programmable logic are as under :
    1. It saves valuable board space or real estate power and time, which reduce the cost.
    2. It increases programmable and design security. There is a security fuse, which can be used to protect proprietary installation properly.
    3. Integrator increases design reliability, because there are fewer dependencies on the interconnection of devices.
    4. Greatest advantage of programmable logic is flexibility.
    5. Programmable logic allows you to use design tools that help to automate the process.
- PLDs have the advantages as under :

1. Short design cycle	2. Design flexibility
3. Low development cost	4. Improved reliability
5. Reduced space	6. Low power consumption
7. High design security	8. Reduced complexity
9. Compact circuitry	10. Field programmable
11. Higher switching speed	12. Reprogrammable

1.13.6 Programming Technologies

- Programmable technologies used for PLDs are as under :

Technology	Predominantly associated with
1. Fusible-link	Simple PLDs (SPLDs)
2. Antifuse	FPGAs
3. EPROM	SPLDs and CPLDs
4. EEPROM and FLASH	SPLDs and CPLDs (some CPLDs)
5. SRAM	FPGAs (some CPLDs)

- EPROM, EEPROM and FLASH technologies are commonly used in PLDs and CPLDs to establish programmable connection.
- FPGAs commonly use SRAM and Antifuse technologies.
- EEPROM and FLASH use electrical erasing, while EPROM requires UV erasing which is time consuming and therefore expensive.
- Electrical erasing technologies are most commonly used.

1.13.7 Applications

1. Gate Logic	2. State Machines
3. Counters	4. Synchronization
5. Decoders	6. Bus Interfaces
7. Parallel-to-Serial register	8. Serial-to-Parallel register

9. Data communication

10. Signal processing

11. Data display

12. Piming and control operation.

1.13.8 Types of PLDs

- PLDs are classified into three categories as under :

    1. Simple Programmable Logic Devices (SPLDs).

    2. Complex Programmable Logic Devices (CPLDs).

    3. Field Programmable Logic Derives (FPLDs).

- Simple Programmable Logic Devices (SPLDs) are also known as (1) PAL (Programmable Array Logic), (2) GAL (Generic Array Logic), (3) PLA (Programmable Array Logic) and (4) PLD (Programmable Logic Devices).

1.14 READ ONLY MEMORY (ROM) AS A PLD

1.14.1 Concept

- A ROM is used for storing programs that are permanently resident in the computer and the tables of constants that do not change in value once the production of the computer is completed.

- Thus, a ROM can store an array of binary data. Data stored in ROM can be read out whenever required, but cannot be changed under normal operating condition.

1.14.2 Basic ROM Structure

- Fig. 1.27 shows a basic structure of ROM that has 'n' input lines and 'm' output lines.

- It maintains an array of 2^n words and each word is 'm' bits long.

- The input line serves as an address to select one of the 2nd words.

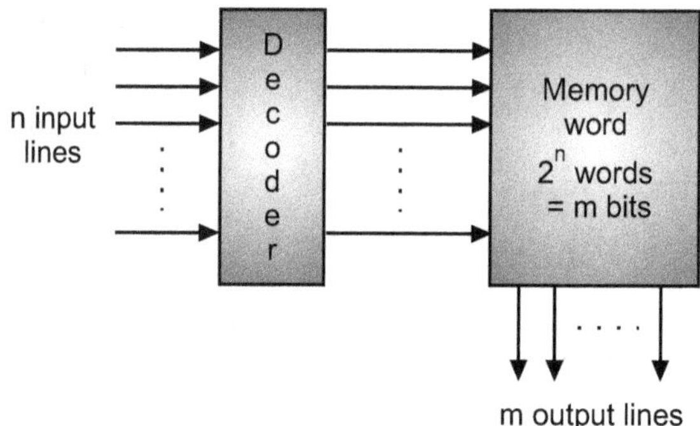

Fig. 1.27 : Basic ROM structure

- A ROM consists of a decoder and memory array.

- When binary input is applied to the decoder inputs, only one of the 2^n decoder outputs is 1, which selects one of the words in the memory array and binary data stored in this word is transferred to the output memory lines.

- A $2^n \times m$ ROM can realize 'm' functions of 'n' variables. Thus, ROM can store a truth table with 2^n rows and m columns.

- Mask programmable ROMs and erasable programmable ROMs called EPROMs are the basic types of ROM.

- Data is permanently stored in a mask programmable ROM at the time of manufacture.

- Use of mask programmable ROMs is economical only if large quantities are required with the same data entry. If only a small quantity of ROMs are required with a given data array EPROMs may be used.

- During the development phase, it is needed to modify the data stored in a ROM, so EPROMs are used instead of mask programmable ROM.

- By using PROM programmer, data stored in EPROM can be changed.

- Data is erased using an ultraviolet (UV) light.

- The electrically erasable PROM (EEPROM) is a more resent development.

- EEPROM is similar to EPROM except that stored data is erased using electrical pulses instead of ultraviolet light.

- Flash memories have built-in programming and erase capability so that data can be written to a flash memory placed in a circuit without the need of separate programmer.

1.14.3 ROM Organization

- Fig. 1.28 shows logic diagram of a 16 bit ROM array.

- To select any one of the 16 bits, a 4 bit address (A_1 A_2 A_3 A_4) is required.

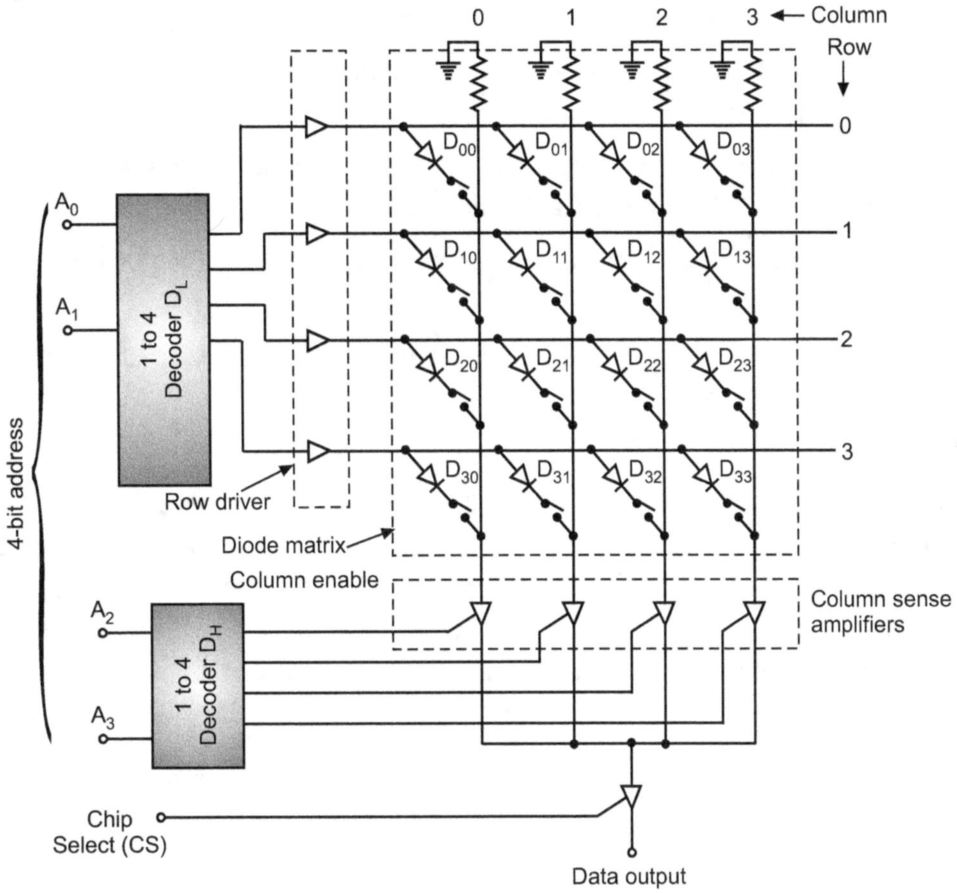

Fig. 1.28 : 16 bit ROM array

- The lower order two bits (A_1A_2) are decoded by the decoder D_L which selects one of the four rows, whereas the higher order two bits (A_3A_4) are decoded by the decoder D_H, which activates one of the four column sense amplifiers.

- The output is enabled by applying logic 1 at the chip select (CS) input.

- Programming a ROM means to selectively open and close the switches in series with the diodes.

- For example, if the switch of diode D_{21} is in closed position and if the address input is 0110, row 2 is activated connecting it to column 1.

- Also the sense amplifier of column 1 is enabled which gives logic 1 output, if the chip is selected (CS = 1). This shows that a logic 1 is stored at the address 0110.

- If the switch of the diode D_{21} is open, logic 0 is stored at the address 0110.

- A ROM consists of an array of BJTs or FETs are also available.

1.15 COMPLEX PROGRAMMABLE LOGIC DEVICES (CPLDs)

1.15.1 Introduction

- A complex programmable logic device is an innovative product as compared to earlier logic devices like programmable logic arrays (PLAs) and programmable array logic (PAL). The earlier logic devices were not programmable, so the logic was built by combining multiple logic chips together.

- A CPLD has a complexity between PALs and field-programmable gate arrays (FPGAs). It also has the architectural features of both PALs and FPGAs. The main architectural difference between a CPLD and FPGA is that FPGAs are based on lookup tables, whereas CPLDs are based on sea-of-gates.

1.15.2 Definition

- The acronym of the CPLD is "Complex Programmable Logic Devices".

- **A CPLD is a one kind of integrated circuit that application designers design to implement digital hardware like mobile phones.**

- The CPLD can handle knowingly higher designs than SPLDs (simple programmable logic devices), but offer less logic than FPGAs (field programmable gate arrays).

- **A CPLD is a programmable logic device with complexity between that of PALs and FPGAs, and architectural features of both.**

- The main building block of the CPLD is a macrocell, which contains logic implementing disjunctive normal form expressions and more specialized logic operations.

1.15.3 Need

- The Programmable Logic Devices (PLDs) such as PLAs and PALs have limited number of inputs, product terms and outputs. These devices can support up to about 32 total number of inputs and outputs only. For implementation of circuits that require more inputs and outputs than that are available in a single SPLD chip, either multiple SPLD chips can be employed. But this has also some limitations. To overcome these limitations, more sophisticated chip like CPLD is needed to design the digital circuits.

Concepts:

- The complexity of any digital IC chip can be specified in terms of number of equivalent 2-input NAND gates.

- A typical PAL has 8 macro cells, if each macro cell represents about 20 equivalent gates, than the PAL can accommodate a circuit that needs up to about 160 gates.

- For a circuit requiring very large number of gates, CPLDs having large number of macro cells (say 512 macro cells) can implement circuits of up to about 10 thousand equivalent gates i.e. CPLDs are similar to SPLDs except that the CPCD is equivalent of 2 to 64 SPLDs.

- A CPLD typically contains from tens to a few hundred macro cells.

- CPLDs are as fast as PLAs but more complex.

- *A Complex Programmable Logic Devices (CPLDs) are the digital IC chip (ASICs) that are just like a large number of PALs in a single silicon chip connected to each other through a cross point switch.*

1.15.4 Features

1. Non-volatile memory
2. Routing
3. Large number of gates
4. More flexible logic

1.15.5 Block Diagram

- Fig. 1.29 shows a block diagram of CPLD.

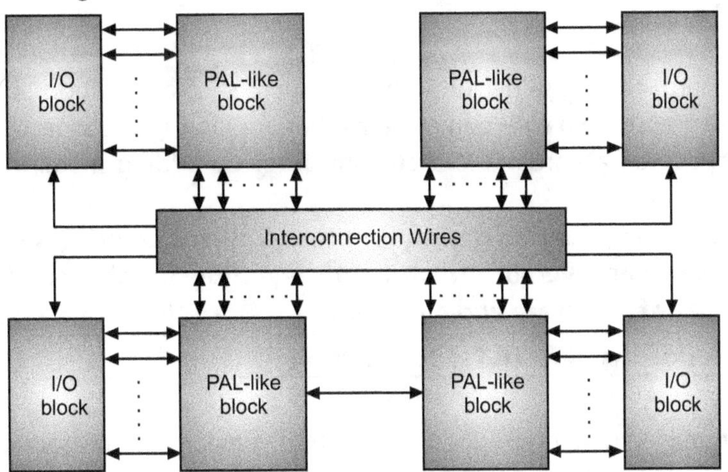

Fig. 1.29 : Block diagram (architecture) of CPLD

- It consists of a number of PAL like blocks, input/output blocks and a set of interconnection wires.

- The PAL like blocks are connected to a set of interconnection wires and to an input/output block.

- The input block is used to drive signals to the pins of the CPLD device at the appropriate voltage levels with the appropriate current.

- A PAL-like block (also called functional block) usually consists of 16 macro cells.

- Each macro cell consists of an AND-OR configuration, an EX-OR gate, a flip-flop, a multiplexer and a tristate buffer.

- Each AND-OR configuration usually consists of 5 to 20 AND gates and an OR gate with 5 to 20 inputs.

- An EX-OR gate is used to obtain the output of OR gate in an inverted or non-inverted form depending upon its other input being 1 or 0 respectively.

- A D-flip-flop stores the output of the EX-OR gate, a multiplexer selects either the output of the D-flip flop or the output of EX-OR gate depending upon the select input (1 or 0).

- The tristate buffer acts as a switch, which enables the chips pin to be used as an output (tristate enabled) or as an input (tristate disabled).

- In case the chip's pin is used as an input pin, an external source can drive a signal on to the pin, which can be connected to other macro cells using the interconnection wiring.

1.15.6 Applications

1. Critical boot loader
2. Digital design
3. Loading the configuration
4. Data of a FPGA
5. Address decoding
6. Cost-sensitive, battery-operated

1.16 GATE ARRAYS

1.16.1 Introduction

- Experts often consider gate arrays as part of a specific kind of application-specific integrated circuits (ASICs) that allow for Printed Circuit Board (PCB) development, along with the development of the hardware and software systems that these components go into.

- In general, this type of chip is a prefabricated chip using NAND or other logical gates, often in a scenario where engineers might include different sets of transistors and resistors that, although built into the circuit board, are not connected for logical operation.

- The first CMOS gate array were developed by **Robert Lipp** in 1974 for International Microcircuits. In 1981, **Wilfred Corrigan**, Bill O'Meara Rob Walker and Mitchell "Mick" Bohn founded LSI Logic to commercialize ECL gate arrays.

- A gate array is a prefabricated silicon chip with most transistors having no predetermined function. CMOS (complementary metal-oxide-semiconductor) gate arrays were later developed and came to dominate the industry.

1.16.2 Definition

- A gate array is an approach to the design and manufacture of application-specific integrated circuits (ASICs) using a prefabricated chip with components that are later interconnected into logic devices (e.g. NAND gates, flip-flops, etc.) according to a custom order by adding metal interconnect layers in the factory.
- **A gate array is a specific engineering design for printed circuit boards (PCBs).**
- **Gate arrays are part of the overall infrastructure equation for the physical circuit boards that power numerous types of hardware and devices.**

1.17 FIELD PROGRAMMABLE GATE ARRAYS (FPGAs)

1.17.1 Introduction

- The FPGA configuration is generally specified using a HDL, similar to that used for ASIC.
- Altera was founded in 1983 and delivered the industry's first reprogrammable logic device in 1984. Ross Freeman and Bernard Vonderschmitt invented the first commercially viable field-programmable gate array in 1985. The 1990s were a period of rapid growth for FPGAs, both in circuit sophistication and the volume of production. In the early 1990s, FPGAs were primarily used in telecommunications and networking.
- FPGAs contain an array of programmable logic blocks, and a hierarchy of "reconfigurable interconnects" that allow the blocks to be "wired together", like many logic gates that can be inter-wired in different configurations. Logic blocks can be configured to perform complex combinational functions, or merely simple logic gates like AND and XOR. In most FPGAs, logic blocks also include memory elements, which may be simple flip-flops or more complete blocks of memory.
- Many FPGAs can be reprogrammed to implement different logic functions, allowing flexible reconfigurable computing as performed in computer software.

1.17.2 Definition

- FPGA is an acronym for Field Programmable Gate Array. **A FPGA is an integrated circuit designed to be configured by a customer or a designer after manufacturing**
- The FPGA configuration is generally specified using a hardware description language(HDL), similar to that used for an application-specific integrated circuit (ASIC). Circuit diagrams were previously used to specify the configuration, but this is increasingly rare due to the advent of electronic design automation tools.
- FPGAs contain an array of programmable logic blocks, and a hierarchy of reconfigurable interconnects.

1.17.3 Architecture

- The general architecture of an FPGA is as shown in Fig. 1.30. But each vendor has its own FPGA architecture.

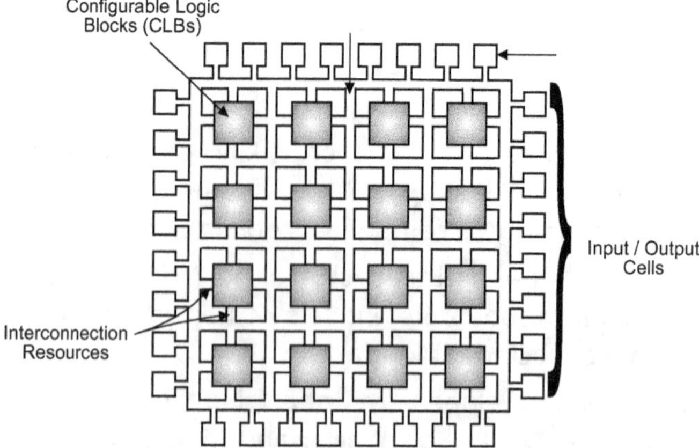

Fig. 1.30 : General architecture of an FPGA

- The architecture of an FPGA consists of configurable logic blocks, configurable input-output blocks and programmable interconnects.
- Also, there will be clock circuitry for driving the clock signals to each logic block and additional logic resources.

1.17.4 Working Principle

- Consider the general architecture of an FPGA as shown in Fig. 1.30.
- It consists of configurable logic blocks, configurable input-output blocks and programmable interconnect. They are as discussed below.

1. Configurable logic blocks :

- Configurable logic blocks (CLBs) contain the logic for the FPGA. In a large grain architecture, these CLBs will contain logic to create a small state machine.
- In a fine grain architecture more like a true gate array ASIC, the CLB will contain only very basic logic.

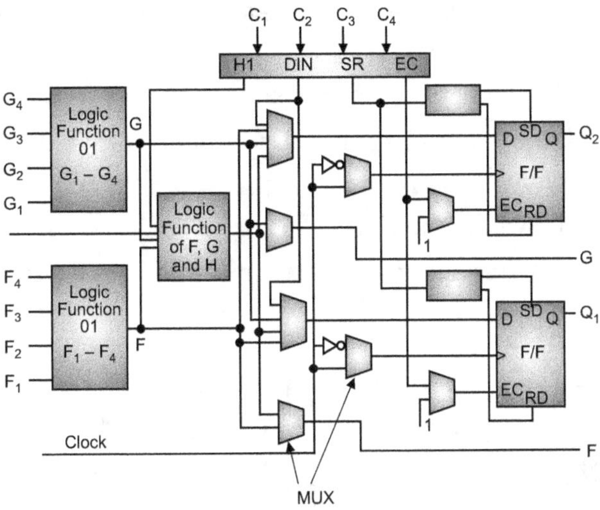

Fig. 1.31 : FPGA configurable logic block

- The logic diagram for FPGA configuration logic block is shown in Fig. 1.31 and it would be considered as a large grain block.
- It contains RAM for creating arbitrary configurable logic functions. It also contains flip-flops for clocked storage elements and multiplexers in order to route logic within the block and to and from external resources.
- The multiplexers also allow polarity selection and reset and clear input relation.

2. Configurable I/O block :

- The logic diagram of a configurable I/O block is shown in Fig. 1.32.

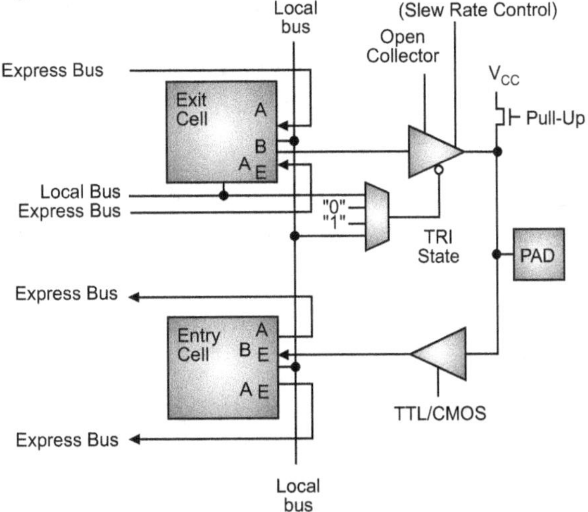

Fig. 1.32 : FPGA configurable input/output block

- It consists of an input buffer and an output buffer with three states, and open collector output controls.

- Typically, there are pull-up resistors on the outputs and sometimes pull-down resistors.

- The polarity of the output can usually be programmed for fast or slow rise and fall times.

- In addition, there is often a flip-flop on outputs so that clocked signals can be output directly to the pins without encountering significant delay.

- It is done for input so that there is no much delay on a signal before reaching a flip-flop which would increase the device hold time requirement.

- A configuration input/output is used to bring the signals into the chip and send them back off again.

3. Programmable interconnect :

- The block diagram of an FPGA programmable interconnect is shown in Fig. 1.33.

- The interconnect of a FPGA is very different than that of a CPLD, but is rather similar to that of a gate array ASIC.

- Fig. 1.33 shows a hierarchy of interconnect resources.

- There are long lines, which can be used to connect critical CLBs that are physically far from each other on the chip without including much delay.

- They can also be used as buses within the chip.

- There are also short lines which are used to connect individual CLBs which are located physically close to each.

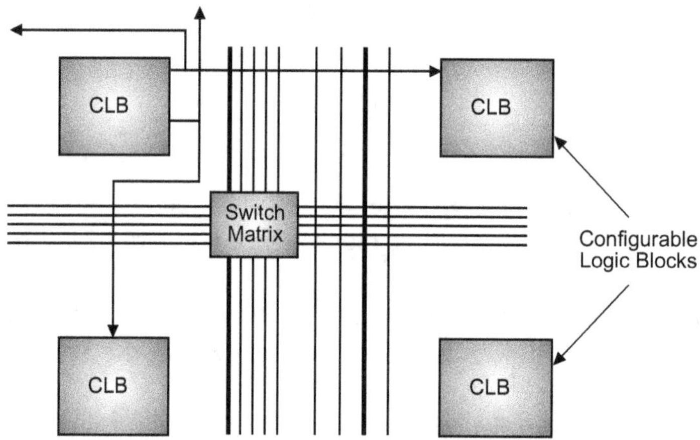

Fig. 1.33 : FPGA programmable interconnect

- There is often one or several switch matrices, like that in a CPLD, to connect these long and short lines together in specific ways.

- Programmable switches inside the chip allow the connection of CLBs to interconnect linear and interconnect lines to each other and to the switch matrix.

- Three state buffers are used to connect many CLBs to a long tine, creating a bus.

- Special long lines called global clock lines are specially designed for low impedance and thus fast propagation times.

- These are connected to clock buffers and to each clocked element in each CLB. This is how the clocks are distributed throughout the FPGA.

1.17.5 Examples of FPGA Families

- The SRAM based FPGA families include the following :

  1. Altera FLEX family.

  2. Atmel AT 6000 and AT 40 K families.

3. Lucent technologies OECA family.

4. Xilinx XC4000, XC5200, Virtex and Sparten families.

- Anti fuse-based FPGA families include the following :

1. Actel PROASIC PLUS families, SX and MX families.

2. Quick logic PASIC family.

1.17.6 Advantages/Benefits

1. High flexibility
2. More acceleration
3. Integration
4. Total cost of ownership
5. Improved performance
6. Reduced costs
7. Quicker time to market
8. More flexible
9. Higher reliability
10. Low maintenance

1.17.7 Factors of Selection

- The selection of FPGA between SRAM or antifuse technology is based on the following factors :

1. Performance of the technology.

2. Density and capacity of that device.

3. Easy to use software.

4. In-system Programmable (ISP) and In-system Reprogrammability (ISR).

1.17.8 Applications

1. Automotive
2. Broadcast
3. Computer and storage
4. Consumer
5. Embedded systems
6. Industries
7. Medical imaging
8. Military
9. Aerospace
10. Government
11. Test and measurement
12. Wireless
13. Wireline
14. Digital singleprocessing (DCP)
15. Bioinformaties
16. Device controller
17. Software defined ratio
18. Random logic
19. ASIC prototyping
20. Computer hare emulator
21. Integrating multiple SPLD
22. Voice recognition,
23. Crypeoloty
24. Filtering and communication electronics,
25. Medical electronics systems
26. Consumer electronics
27. Distributed monetary system
28. Data counter
29. High performance computing
30. Industrial medical
31. Scientific instruments
32. Security system
33. Wireless communications
34. Wired communications.

Practice Questions

1. What is a sequential logic circuit ?

2. What is a sequential logic circuit ?

3. What are the characteristics of sequential logic ?

4. List the examples of sequential logic circuits ?

5. Draw the block diagram of a sequential logic.

6. Define the following terms : (a) Metastability, (b) Noise margin, (c) Power dissipation, (d) Fan-out, and (e) Skew.

7. What does word 'finite' signify in the term FSM ?
8. Define the term state diagram.
9. Define the terms : (a) Sequential state machine, and (b) FSM.
10. Draw state diagram and block diagram of Mealy state machines.
11. Explain the difference between Mealy and Moore types of state machines.
12. Draw state diagram and block diagram of Moore state machine.
13. What are the advantages and demerits of describing state machines with the state diagram compared to ASM, truth table and K-map ?
14. Differentiate between asynchronous and synchronous sequential machines.
15. What is ASIC ?
16. Write a short note on ASIC.
17. Give the classification of ASIC.
18. What is meant by configuration as referred to PLD ?
19. Give classification of semi-custom ASIC.
20. What is PLD ?
21. Why do you need PLD digital design ?
22. What are the features of PLDs ?
23. List the advantages of programmable logic.
24. State advantages of PLDs.
25. State programmable technologies.
26. List the types of PLDs.
27. What is meant by CPLD ?
28. What is the need of CPLD in digital circuits ?
29. Draw architectural diagram of CPLD.
30. Write short note on CPLD architecture.
31. What is a function of PTOF in CPLD ?
32. Give comparison of PLD and ASIC.
33. Give comparison of ASIC and FPGA.
34. Give comparison of CPLD and FPGA.
35. Give comparison of SRAM and Antifuse FPGA.

☆☆☆

CMOS TECHNOLOGY CONCEPTS

Learning Objectives

(2a) Compare the performance of BJT and CMOS for the given parameters.

(2b) Draw the simplified CMOS logic of the given gates.

(2c) Explain CMOS inverter characteristics with relevant sketch.

(2d) Describe the given MOS fabrication process.

2.1 BJT PARAMETERS

2.1.1 Introduction

- There are many things about transistors that confuse the beginner and no beginner alike. Some circuits are easy and do not require much more than Ohm's law, while others seem of great deal harder.
- Paradoxically, it is often the circuit that appear to be the simpler that cause the most problems. A perfect example is a BJT amplifier circuit, using only a single transistor and a pair of resistors. While this topology is easily beaten by even the most op-amp for most things, it offers a fairly easy way to determine the transistor's parameters.
- There are even applications where it is useful, particularly, where there are no op-amps in the circuit and you need a gain stage.

2.1.2 Parameters

1. Determining characteristics
2. Transistor matching
3. AC performance
4. Measured results
5. Stabilizing voltage gain
6. Early effect

1. Determining Characteristics:

- For the time being, we shall ignore the AC performance, and just examine the biasing requirements. The circuit of Fig. 2.1 would appear to be fairly easy to analyse, because it is so simple.
- The analysis problem lies in the work *'feedback'*. Whatever happens at the collector is reflected back to the base, so the collector voltage is dependent on the base current, which in turn is dependent on the collector voltage.
- The transistor h_{fc} changes the relationship between collector and base, and without knowing one of the parameters in advance, it is simply not possible to predict exactly what the circuit will do.

Current gain, $$\beta = \frac{R_1}{R_2} \qquad \dots (2.1)$$

- Assume that β of a transistor Q is 240. That means its base current is $\dfrac{I_c}{240}$. If since there is 6V across the resistor R_1, the current must be $\dfrac{6}{R_1}$ mA.

  That means the base current can be estimated at 25 mA, if R_1 = 1 kΩ. The voltage across the resistor R_2 would be V = $I_C R_2$ = 6V, if R_2 = 240 K. The important thing is that if the circuit is designed properly, it will work intended almost regardless of the transistor used.

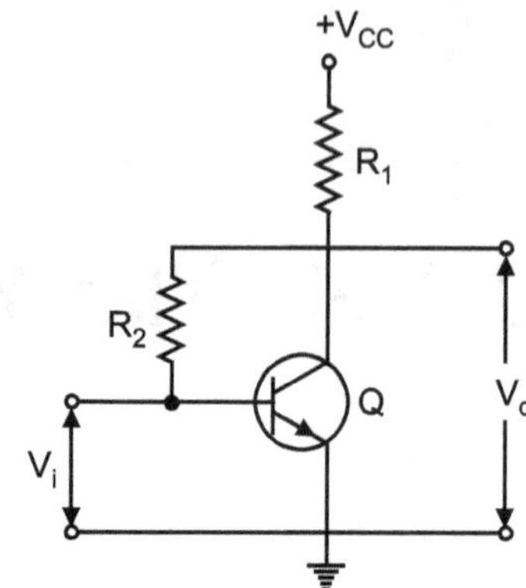

Fig. 2.1 : Common base feedback bias BJT amplifer

2. Transistor Matching:

- The thing that needs to be determined first is the expected collector current, and knowing the common base voltage that will apply in the circuit no extra matched devices may also help. However, you do not really need to provide the full common base voltage that will ultimately be used.

- Current through each transistor should be equal, e.g. 2 mA. A 20 V supply will do just fine for most tests, and good results can still be obtained with a lower voltage. Based on transistor datasheet, you can get a reasonable initial estimate of h_{fe} and use a collector resistor R_1 that will drop about 2 V at 2 mA.

- Then select an appropriate common base resistor R_2, check and use a 1 mΩ resistor in series with a 1 MΩ potentiometer. The transistor should be installed into three receptacles I_C of an I_C socket or use a solderless breadboard.

- Once you have a potentiometer setting that drops 2V across the 1kΩ resistor R_1, the current I_C is 2 mA. Then just install transistors until you find a pair that have the same voltage drop across the collector or resistor R_1, and the same base-emitter voltage. There will inevitably be a small discrepancy because finding two that are identical is unlikely, but if they are within 5% of each other, then that is perfectly acceptable. When installed in the PCB, the two transistors should be thermally bonded, and that ensures that thermal changes affect both devices equally.

3. AC Performance:

- In a simulation with three different transistor types, the AC output voltage is 161 mV, 170 mV and 132 mV for an input of 1 mV from a 50 Ω source.

- The variation from the highest to lowest gain is only a fraction over 2 dB, and these are very different devices. It is educational to look at their datasheets to see just how different they are, yet all work nearly as well as the other without changing the circuit.

- Note that a single-stage amplifier such as this is inverting, and it makes no difference, if you use a valve, BJT, JFET or MOSFET. When operated with the emitter, cathode or source grounded, all devices are inverting. A positive-going input causes a negative-going output and vice versa.

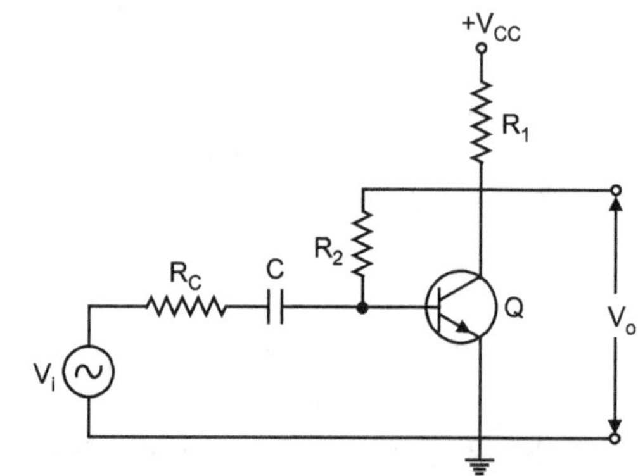

Fig. 2.2 : Common base feedback bias (AC Measurement)

- In Fig. 2.2, common base resistor R_2 is feedback resistor.

- The feedback is negative, so if the collector voltage attempts to rise, more base current is available (via resistor R_2) and the transistor Q is turned ON to keep the collector voltage stable. This feedback acts on both AC and DC signals, and the input impedance is very low. It is useful only for low impedance sources such as audio applications.

4. Measured Results:

- In Fig. 2.2, the supply voltage V_{CC} is set to +12V, and more number of transistors are used. Most were BC546 types (only 4 test results are shown), but from two different makers, and also tested a few BC550C devices as well.

- Also 6 tested a BC550C with emitter and collector reversed. The measured results are shown in Table 2.1. The voltage V_{DC} is about 680 mV for BC546 tests, but it was not measured for BC550s.

Table 2.1

V_{CE} (DC)	h_{FE} (Calculated)	AC Output (RMS)	AC Gain
1. For BC546			
5.94 V	276	1.52 V	152
5.80 V	291	1.52 V	152
5.77 V	293	1.52 V	152
7.20 V	177	1.24 V	124 (– 1.8 dB)
2. For BC550C			
4.36 V	498	2.04 V	204
4.31 V	508	2.04 V	204
3.65 V	675	2.12 V	212 (+ 0.3 dB)
4.83 V	414	1.92 V	192 (– 0.5 dB)
3. For BC550C Reversed			
8.39 V	112	255 mV	25.5

- When a low collector voltage is measured, the transistor has a high DC gain and vice versa. Equally, several transistors show identical voltage gain, even when it is apparent that their h_{FE} is different but h_{FE}, and B gains are very close.

- Match transistors as described in point (2) if that is essential for the circuit you are building. Note that the B-E voltage V_{be} is still a variable, and that needs to be matched independently of h_{FE}.

5. Stabilising Voltage Gain:

- In most cases, a defined gain is required, and that is achieved with the addition of another resistor. In the drawing below, I added a 100 Ω emitter resistor. We have added a 100 Ω another resistor R_E to Fig. 2.3. The gain is now determined by the ratio of R_1 to R_3.

- With 100 ohms as shown in Fig. 2.3, the theoretical gain is about 9.57 but it does not quite make it because the transistor has infinite gain, so the feedback cannot produce an accurate result. However, it is too bad, and far more predictable than you would expect otherwise.

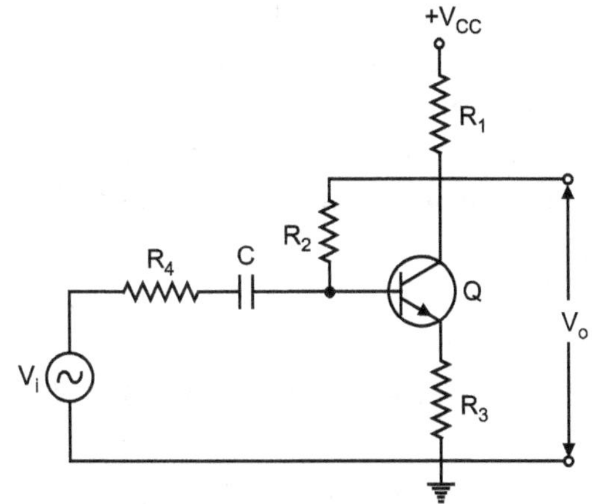

Fig. 2.3 : Common Base Amplifier with emitter resistor

- Ideally Fig. 2.3 would have been rebiased to get a collector voltage close to 6.5 V across the emitter resistor R_E. However, even with the same batch of very different transistors, the gain variation between the highest and lowest gain is now a mere 0.15 dB.

- Distortion is also reduced, but *not* by the same ratio as the gain reduction. The addition of an emitter resistor is called emitter degeneration, and it is not the same thing as negative feedback. It is effective for stabilising the gain, but does not reduce distortion as well as 'true' negative feedback.
- The noise from the resistor R_3 is actually amplified by this circuit and all similar arrangements, so despite the reduction of gain, the noise will not be reduced in proportion. The gain is reduced further when an external load is added, because that is effectively in parallel with the collector resistor R_1.

6. Early Effect:

- The *Early effect* is named after its discoverer, *James Early*. It is caused by the variation in the *effective* width of the base in a BJT, due to a change of the applied base to collector voltage V_{bc}. Remember that in normal operation, the base-collector junction is reverse biased, so a *greater* reverse bias across this junction increases the collector–base depletion width. This decreases the width of the charge carrier portion of the base, and the gain of the transistor is increased. The transistor's Early effect has some influence over the performance for AC and DC.
- As you can see from the graph, the slope is quite linear. It follows that as the collector voltage changes, so does the effective h_{FE}.
- There is also a change of r_e as the collector current I_C varies, it becomes relevant only when *voltage* gain is expected. No collector load resistor R_C is used because the base current is maintained at a constant and very low value. Over the full range shown below, the AC gain changes by a factor of about 1.6:1 for the current range seen in Fig. 2.3, and with a collector voltage between 1 and 50 V.
- The AC voltage gain is almost directly proportional to the collector current. Voltage gain is not relevant to this test because only current is monitored. In more complex circuits, it is common to keep the transistor's collector voltage as constant as possible.

2.2 CMOS PARAMETERS

2.2.1 Introduction

- **Complementary Metal–Oxide–Semiconductor (CMOS) is a type of MOSFET fabrication process that uses complementary and symmetrical pairs of p-type and n-type MOSFETs for logic functions.**
- CMOS technology is used for constructing ICs. It is also used for analog and digital circuits.
- The MOSFET or MOS transistor was invented by Mohamed M. Atalla and Dawon Kahng at Laboratories in 1959.

2.2.2 General Test Parameters

1. Logic high input voltage,	2. Logic low input voltage,
3. Logic high output voltage,	4. Logic low output voltage,
5. Logic high input current,	6. Logic low input current,
7. Logic high output current,	8. Logic low output current

2.2.3 Fundamental Parameters

1. Area	2. Complexity
3. Performance	4. Power consumption
5. Robustness	6. Reliability

2.3 COMPARISON OF BJT AND CMOS TECHNOLOGIES

Table 2.2 : Comparison of CMOS and BJT Technologies

Bipolar Technology	CMOS Technology
1. It has high power dissipation.	1. It has low static power dissipation.
2. It has high output drive current.	2. It has low output drive current.
3. It has low input impedance.	3. It has high input impedance.

Contd...

4. It has low packing density.	4. It has high packing density.
5. It has low delay sensitivity to load.	5. It has high delay sensitivity to load.
6. It has more fan-out.	6. It has less fan-out.
7. It has unidirectional capability.	7. It has bidirectional capability.
8. Collector and emitter cannot be interchanged.	8. Drain and source can be interchanged.
9. It is not an ideal switching device.	9. It is an ideal switching device.

2.4 CMOS LOGIC GATES

2.4.1 Introduction

- In CMOS logic gates a collection of NMOS is arranged in a pull-down network between the output and the low voltage power supply rail (V_{ss} or quite often ground).

- Instead of the load resistor of NMOS logic gates, CMOS logic gates have a collection of p-type MOSFETs in a pull-up network between the output and the higher-voltage rail (often named V_{DD}).

- Thus, if both NMOS and PMOS have their gates connected to the same input, the PMOS will be ON when the NMOS is OFF, and vice-versa. The networks are arranged such that one is ON and the other is OFF for any input pattern as shown in Fig. 2.4.

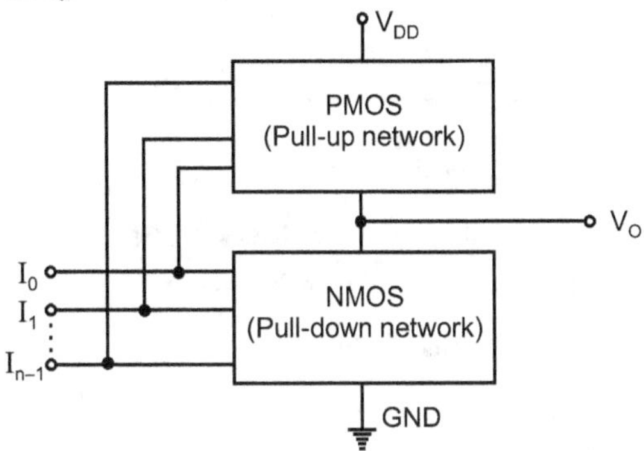

Fig. 2.4 : CMOS logic gates using pull-up and pull-down networks

- CMOS offers relatively high speed, low power dissipation, high noise margins in both states, and will operate over a wide range of source and input voltages provided the source voltage is fixed.

2.4.2 CMOS Inverter

- The CMOS inverter circuit is shown in Fig. 2.5. It consists of NMOS and PMOS transistors. The input A serves as the gate voltage for both NMOS and PMOS transistors.

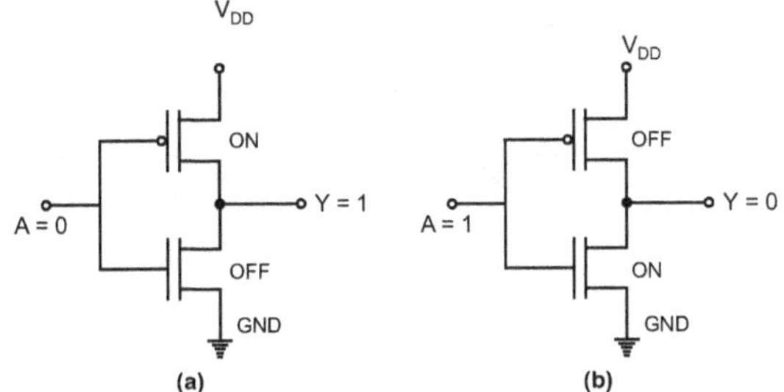

Fig. 2.5 : CMOS inverter

Table 2.3

Input	Output
0	1
1	0

- The NMOS transistor has an input from V_{ss} (ground) and PMOS transistor has an input from V_{DD}. The terminal is an output.

- When a high voltage ($\sim V_{DD}$) is given at input terminal (A) of the inverter, the PMOS switched off circuit and NMOS switched on so the output Y will be pulled down to V_{ss}.

- When a low-level voltage ($< V_{DD}$, ~ 0 V) applied at input terminal A of the inverter, the NMOS switched OFF and PMOS switched ON. So the output Y is pulled up to V_{DD}.

- The working of CMOS is summarized in Table 2.4.

Table 2.4

Input	Logic Input	Output	Logic Output
0 V	0	V_{DD}	1
V_{DD}	1	0 V	0

2.4.3 CMOS OR Gate

- Fig. 2.6 shows the logic circuit of an OR gate using CMOS. It consists of CMOS NOR and NOT gates using CMOS.

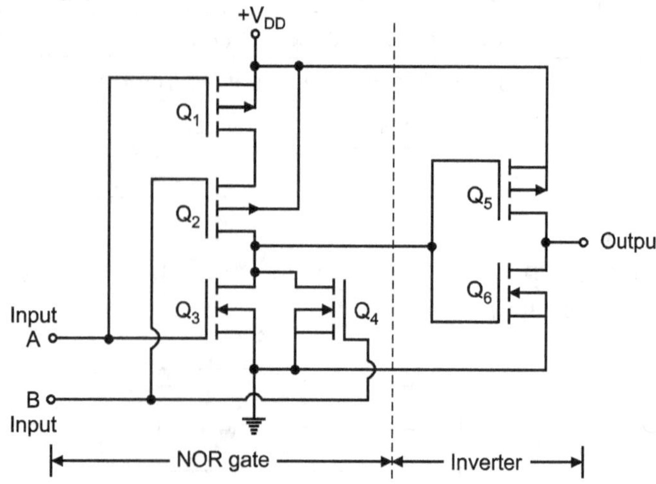

Fig. 2.6 : CMOS OR Gate

Table 2.5		
Inputs		Output
A	B	
0	0	0
1	0	1
0	1	1
1	1	1

- Truth table of an OR gate is shown in Table 2.5.

2.4.4 CMOS AND Gate

- Fig. 2.7 shows the logic circuit of an AND gate using CMOS. It consists of CMOS AND and NOT gates using CMOS.

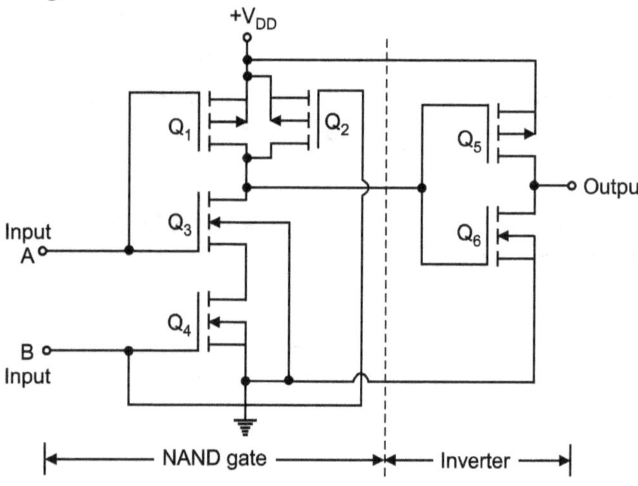

Fig. 2.7 : CMOS AND Gate

Table 2.6		
Inputs		Output
A	B	
0	0	0
1	0	0
0	1	0
1	1	1

- Truth table of an AND gate is shown in Table 2.6.

2.4.5 CMOS NOR Gate

- A 2-input CMOS NOR gate is shown in Fig. 2.8. It consists of two parallel NMOS transistors between output Y and ground, while two series PMOS transistors between output Y and V_{DD}.

- The NMOS transistors are in parallel to pull-down the output Y low, when either input A or B is high.

- The PMOS transistors are in series to pull-up the output Y high, when both inputs A and B are low. The output is never floating.

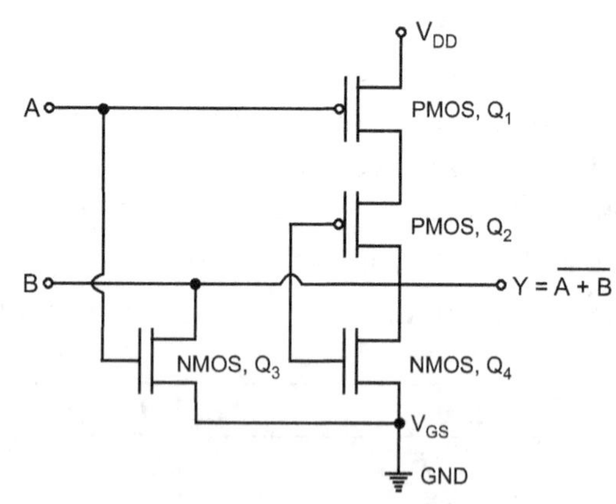

Fig. 2.8 : CMOS NOR gate

Table 2.7

Inputs		State of CMOS				Output
A	B	Q_1	Q_2	Q_3	Q_4	Y
0	0	1	1	0	0	1
0	1	1	0	1	0	0
1	0	0	1	1	0	0
1	1	0	0	1	1	0

- The truth table of CMOS NOR gate is shown in Table 2.7.

2.4.6 CMOS NAND Gate

- Fig. 2.9 shows a 2-input complementary MOS NAND gate. It consists of two series NMOS transistors between Y and ground and two parallel PMOS transistors between output Y and V_{DD}.

- If either input A or B is logic 0, at least one of the NMOS transistors will be OFF, breaking the path from Y to ground (GND). But at least one of the PMOS transistors will be ON, creating a path from Y to V_{DD}.

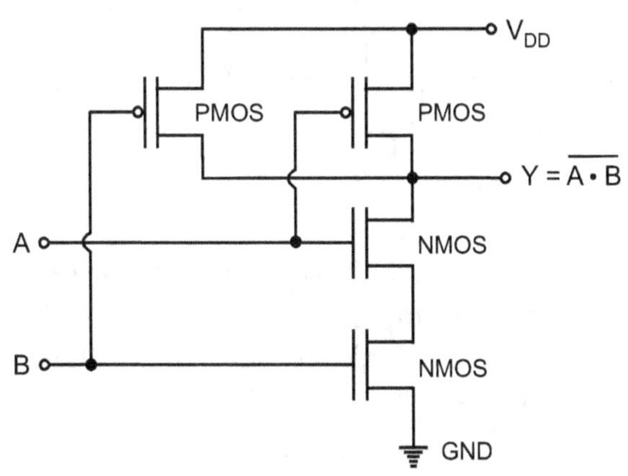

Fig. 2.9 : CMOS NAND gate

Table 2.8

Inputs		Pull-Down Network	Pull-Up Network	Output Y
A	B			
0	0	OFF	ON	1
0	1	OFF	ON	1
1	0	OFF	ON	1
1	1	ON	OFF	0

- Hence, the output Y will be high. If both inputs A and B are high, both of the NMOS transistors will be ON and both of the PMOS transistors will be OFF. Hence, the output will be logic low. The truth table of NAND logic gate given is shown in Table 2.8.

2.4.7 CMOS Transmission Gate

- *A transmission gate is an analog gate similar to an electromagnetic relay that can conduct in both directions or block a control signal with almost any voltage potential.*

- **A digitally controlled CMOS switch is called transmission gate.** If the CMOS gate transmits the signal in both directions, then it is called a **bilateral transmission gate**. It is also known as **bilateral switch**.

- A transmission gate is a bilateral switch consisting of NMOS and PMOS transistors controlled by externally applied logic levels.

- The enhancement MOSFET can be used as an analog switch. This switching is purely resistive. An NMOS transistor behaves like an closed (ON) switch, when the gate-source voltage, V_{GS} is greater than the threshold voltage V_{TH}. Also it behaves like an open (OFF) switch, when the gate source voltage, V_{GS} is less than the threshold voltage V_{TH}.

- A PMOS transistor behaves like a closed (ON) switch, when the gate-drain voltage, V_{GD} is less than the threshold voltage, V_{TH}. Also it behaves like an open (OFF) switch, when the gate-drain voltage V_{GD} is greater than the threshold voltage V_{TH}. Thus the gate-source voltage, V_{GS} act as an ON/OFF control voltage for eMOSFETs.

- The NMOS and PMOS transistors are connected back to back in parallel with an inverter I used between the gate of the NMOS and PMOS to provide the two complimentary control voltages, V_c and $\overline{V_c}$.

- When the input control signal V_c is Low (Logic '0') then both the NMOS and PMOS transistors cut-off (Logic '0') and the switch is open (OFF). When the input control signal V_c is High (Logic '1') then both the NMOS and PMOS transistors are biased into conduction and the switch is closed (ON).

- Thus, the transmission gate acts as a closed (ON) switch, when $V_C = 1$, while it acts as an open switch (OFF), when $V_C = 0$. Thus the transmission gate operates as a voltage controlled switch.

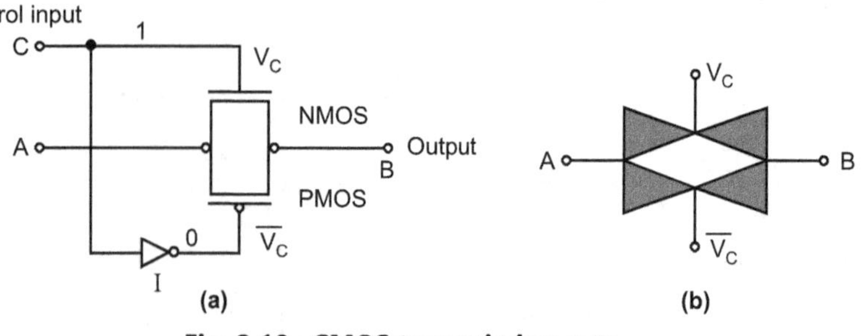

Truth Table 2.9

Control V_c	Input A	Output B
1	0	0
1	1	1
0	0	Hi-z
0	1	Hi-z

Fig. 2.10 : CMOS transmission gate

- We can see from the Truth Table 2.9 that the output at B relies not only on the logic level of the input A, but also on the logic level present on the control input V_C. Thus the logic level value of B is defined as both input A AND control V_C giving us the Boolean expression for a transmission gate of $B = A - V_C$.

- Since the Boolean expression of transmission gate incorporates the logical AND function, it is therefore possible to implement the AND operation using a standard 2-input AND gate with one input being the data input, while the other is the control.

- The transmission gates are used in order to implement electronic switches, analog multiplexers and logic circuits.

2.4.8 MOS Transistor Switches

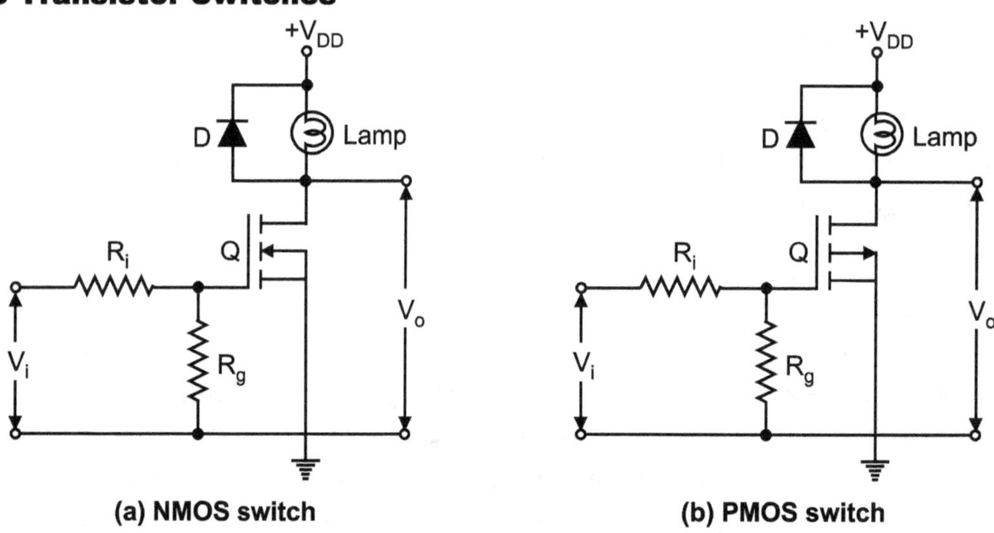

(a) NMOS switch   **(b) PMOS switch**

Fig. 2.11 : MOS Transistor Switches

- MOS transistor can be used to switch a simple lamp ON and OFF. Fig. 2.11 (a) shows the circuit diagram of an NMOS transistor switch.

- The gate input voltage V_{GS} (= V_i) is taken to an appropriate positive voltage level to turn the NMOS transistor and therefore the lamp load L either ON, when voltage V_{GS} is positive voltage level and turns the NMOS OFF, when the voltage V_{GS} is zero or less.

- Fig. 2.11 (b) shows the circuit diagram of PMOS transistor switch. The gate input voltage V_{GS} (= V_i) is taken to an appropriate negative voltage level to turn the NMOS transistor and therefore the lamp load L either ON, when voltage V_{GS} is negative voltage level and turns the MOSFET OFF, when the voltage V_{GS} is positive.

- We can summarize the switching characteristics of both the N-channel and P-channel type MOSFET within the following table.

Table 2.10

MOSFET Type	$V_{GS} \ll 0$	$V_{GS} = 0$	$V_{GS} \gg 0$
1. N-channel Enhancement	OFF	OFF	ON
2. N-channel Depletion	OFF	ON	ON
3. P-channel Enhancement	ON	OFF	OFF
4. P-channel Depletion	ON	ON	OFF

2.4.9 CMOS Inverter Characteristics

- Fig. 2.12 shows the voltage transfer characteristics of a CMOS inverter. Considering the static condition first, in region 1 for which V_{in} = logic 0, the PMOS transistor is fully turned OFF. Thus no current flows through the CMOS inverter and the output is directly connected to V_{DD} through the PMOS transistor.

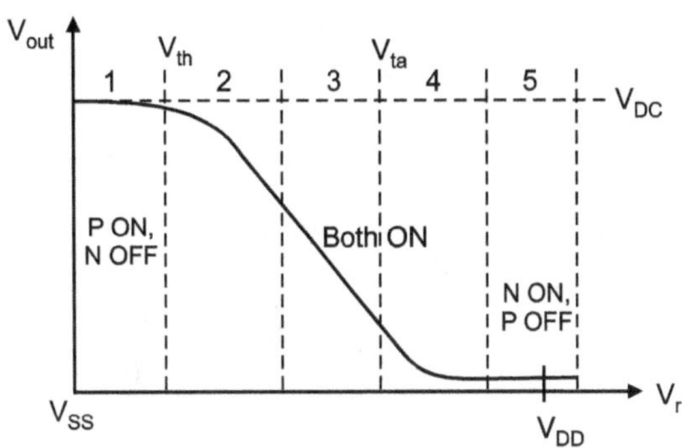

Fig. 2.12 : CMOS inverter voltage transfer characteristics

- In region 5 V_{in} = logic 1, the NMOS transistor is fully ON, while the PMOS transistor is fully turn OFF. Again no current flows and a good logic 0 appears at the output.

- The region 2 the input voltage has increased to a level which just exceeds the threshold voltage of the NMOS transistor. The NMOS transistor conducts and has a large voltage between source and drain. The PMOS transistor also conducting but with only a small voltage across it, it operates in the unsaturated resistive region.

- The region 4 is similar to region 2 but with the roles of the PMOS and NMOS transistors reversed. The current magnitude in region 4 and 2 are small and most of the energy consumed in switching from one state to the other is due to the large current which flows in region 2.

- The region 3 is the region in which the inverter exhibits gain and in which both PMOS and NMOS transistors are in saturation. They act as current sources.

2.5 COMPLEX CMOS LOGIC GATES

2.5.1 Introduction

- The realization of complex Boolean functions, which may include several input variables and several product terms, typically require a series-parallel network of NMOS transistors which constitute the so called pull-down network, and a corresponding dual network of PMOS transistors, which constitute the pull-up network.

- A complex CMOS NOR and CMOS NAND gates with 3-input variables can be realized.

2.5.2 3-Input CMOS NAND Gate

- Fig. 2.13 shows a 3-input CMOS NAND gate. It consists of three series NMOS transistors connected between output Y and ground GND, and three parallel PMOS transistors connected between output Y and V_{DD}.

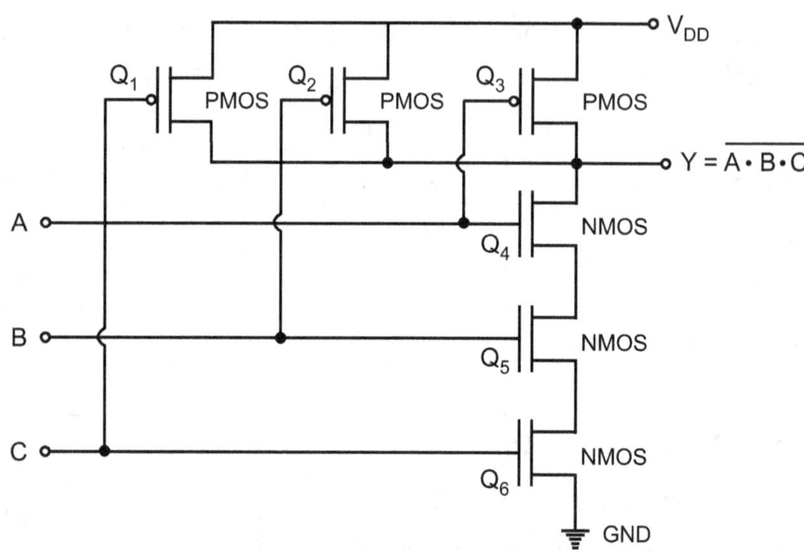

Fig. 2.13 : 3-input CMOS logic gate

Truth Table 2.11

Inputs			State of CMOS						Output
A	**B**	**C**	**Q_1**	**Q_2**	**Q_3**	**Q_4**	**Q_5**	**Q_6**	**Y**
0	0	0	0	0	0	1	1	1	1
0	0	1	1	0	0	1	1	0	1
0	1	0	0	0	1	1	0	1	1
0	1	1	1	1	0	1	0	0	1
1	0	0	0	0	1	0	1	1	1
1	0	1	1	0	1	0	1	0	1
1	1	0	0	1	1	0	0	1	1
1	1	1	1	1	1	0	0	0	1

- The operation of a 3-input CMOS NAND gate is summarized in a Truth Table 2.11.

2.5.3 3-Input CMOS NOR Gate

- A 3-input CMOS NOR gate is shown in Fig. 2.14. It consists of three series NMOS transistors connected between output Y and ground GND, and three series PMOS transistors connected between output Y and V_{DD}.

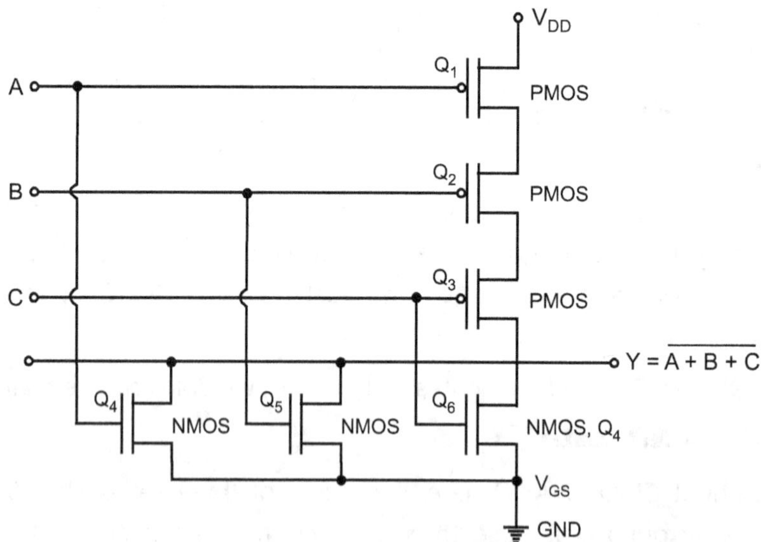

Fig. 2.14 : 3-input CMOS NOR gate

Truth Table 2.12

Inputs			State of CMOS logic						Output
A	B	C	Q_1	Q_2	Q_3	Q_4	Q_5	Q_6	Y
0	0	0	0	0	0	1	1	1	1
0	0	1	0	0	0	0	1	1	0
0	1	0	0	0	1	1	0	1	0
0	1	1	0	1	1	1	0	0	0
1	0	0	1	0	0	0	1	0	0
1	0	1	1	0	0	0	1	0	0
1	1	0	1	1	1	0	0	1	0
1	1	1	1	1	1	0	0	0	0

- The working of a 3-input CMOS NOR gate is summarized in a Truth Table 2.12.

2.5.4 2-Input CMOS AND Gate

- We realize the CMOS AND operation by connecting two NMOS transistors in series.

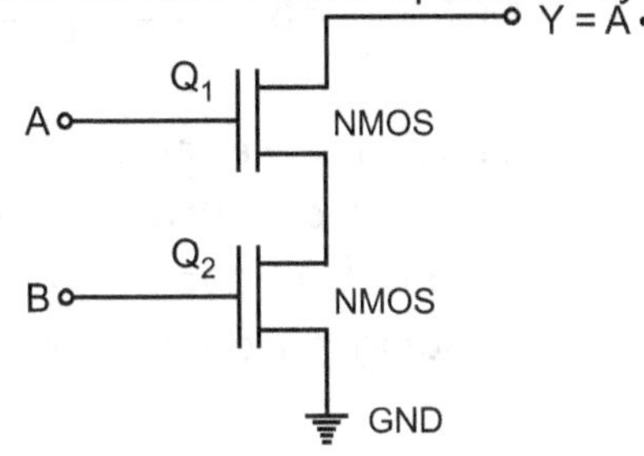

Fig. 2.15 : CMOS AND gate

Truth Table 2.13

Inputs		States		Output
A	B	Q_1	Q_2	Y
0	0	0	0	0
0	1	1	0	0
1	0	0	1	0
1	1	1	1	1

- The truth table of CMOS AND gate is shown in Table 2.13.

2.5.5 2-Input CMOS OR Gate

- CMOS OR gate is realized by connecting two PMOS transistors in parallel.

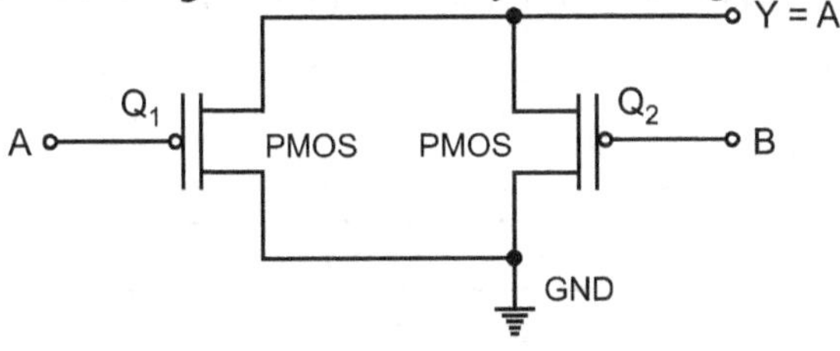

Fig. 2.16 : CMOS OR gate

Truth Table 2.14

Inputs		States		Output
A	B	Q_1	Q_2	Y
0	0	0	0	0
0	1	1	0	1
1	0	0	1	1
1	1	1	1	1

- The truth table of CMOS OR gate is shown in Table 2.14.

2.5.6 3-Input CMOS AND Gate

- Three input variables CMOS AND gate can be realized by connecting three NMOS transistors in series.

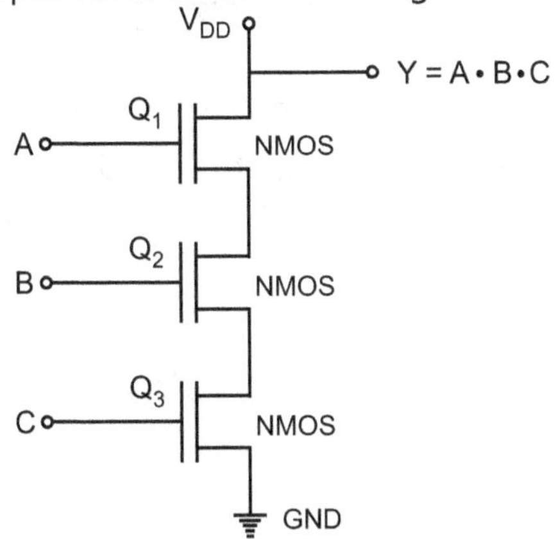

Fig. 2.17 : 3-input CMOS AND gate

Truth Table 2.15

Inputs			States			Output
A	B	C	Q_1	Q_2	Q_3	Y
0	0	0	1	1	1	0
0	0	1	1	1	0	0
0	1	0	1	0	1	0
0	1	1	1	0	0	0
1	0	0	0	1	1	0
1	0	1	0	1	0	0
1	1	0	0	0	1	0
1	1	1	0	0	0	1

- The working is summarized in a Truth Table 2.15.

2.5.7 3-Input CMOS OR Gate

- Three input variables CMOS OR gate is realized by connecting three NMOS transistors in parallel.

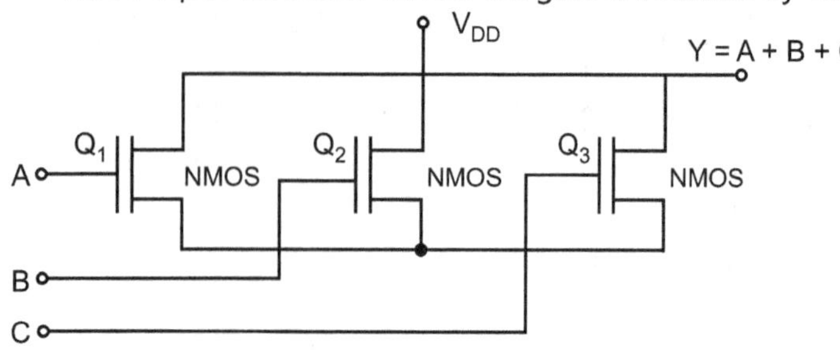

Fig. 2.18 : 3-input CMOS OR gate

Truth Table 2.16

Inputs			States			Output
A	B	C	Q_1	Q_2	Q_3	Y
0	0	0	1	1	1	0
0	0	1	1	1	0	1
0	1	0	1	0	1	1
0	1	1	1	0	0	1
1	0	0	0	1	1	1
1	0	1	0	1	0	1
1	1	0	0	0	1	1
1	1	1	0	0	0	1

- The working is summarized in a Truth Table 2.16.

2.6 RESISTOR AND CAPACITOR LAYOUT

2.6.1 Resistor Layout

- An integrated resistor is made of a strip of a resistive layer. The ending resistance can be significant. The conductive layer in polysilicon resistors, which can be used to shield the conductor-oxide-conductor structure, well or pinched-well layers have a large specific resistance, but they have a large voltage and temperature coefficient. They are weakly insulated from the surrounding. Layers close to the surface contribute to the conductivity.

- In order to use large value resistors, use of long strips (large L/W) and use of layers with high sheet resistance (bad performances) are made.

- Rectangular 'Snake' layout is used for large value resistors. It is not used in precise applications.

- If the parameters are statistically independent, the standard deviation of the resistance is

$$\frac{\Delta L}{L} << \frac{\Delta W}{W}$$

- Since in general L >> W, $\dfrac{\overline{\Delta f}}{\overline{f}}$ for polysilicon resistors is larger than for diffused resistors. Polysilicon is composed of a conglomerate of independently oriented grain of crystalline silicon. Absolute accuracy is poor because of the large parameter drift. Ratio or matching accuracy is better because it depends on the local variation of parameters.

Fig. 2.19 : Rectangular Snake layout for resistors

- Plastic packages cause a large pressure on the die. It determines a variation of the resistivity. The temperature gradient on the chip may produce thermal induced mismatch. The layouts are shown in Fig. 2.19.

- Rules for resistance matching are as under:

  1. Use the same material
  2. Identical geometry some orientation
  3. Close proximity
  4. Interdigitate
  5. Use dummy elements
  6. Use proper endings
  7. Use electrostatic shielding
  8. Place resistors in low stress area
  9. Place resistors away from power devices

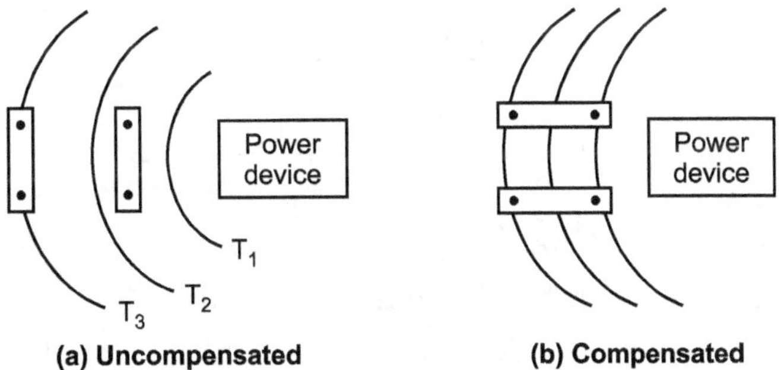

(a) Uncompensated (b) Compensated

Fig. 2.20 : Resistor layout in plastic package

2.6.2 Capacitor Layout

- Intergraded capacitors in IC are parallel-plate capacitors. For integrated capacitors, flux-capacitor layout is used. The features are as under:
    - Use of the same metal layer.
    - Exploit the lateral flux.
    - The parasitic capacitance plate-substrate is low because the metal sits on thick oxide.
    - Use thick metal layers.
    - Maximize the perimeter (use of fractals).
    - Very good matching.
- Fig. 2.21 shows the common controid structure layout used for capacitors. The undercut effect gives the same proportional reduction if the perimeter-area ratio is kept constant.

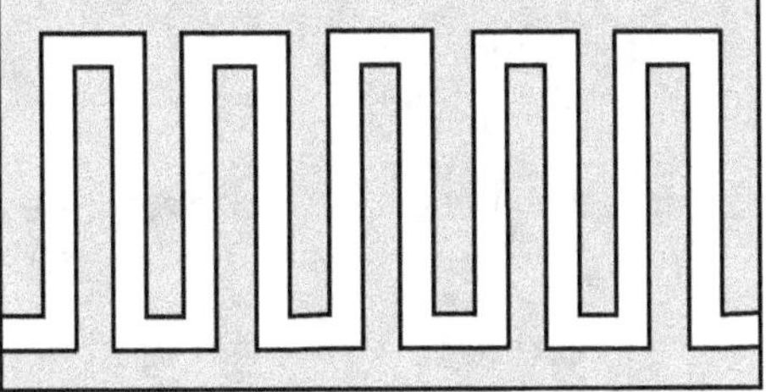

Fig. 2.21 : Fuse capacitor layout

- The rules for capacitor matching are as under:

1. Use identical geometries.
2. Use large unity capacitance (minimize fringing).
3. Use common centroid arrangement.
4. Use dummy capacitors.
5. Use shielding.
6. Account for the connections' contribution.
7. Don't run connections over capacitor.
8. Place capacitor in low stress areas.
9. Place capacitors far from power devices.

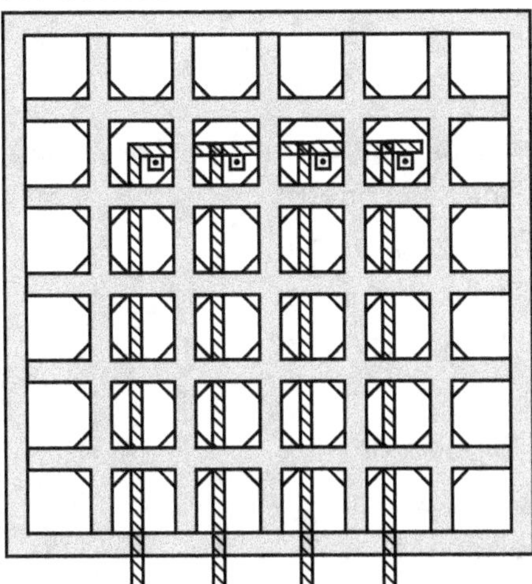

Fig. 2.22 : Common centroid structure layout

2.7 IC FABRIC PROCESS TECHNOLOGY

2.7.1 Introduction

- An Integrated Circuit (IC) is also called as *chip* or *microchip*. It is a semiconductor wafer in which millions of components are fabricated.

- The active and passive components such as resistors, diodes, transistors etc. and external connections are usually fabricated in one extremely tiny single chip of silicon.

- All circuit components and interconnections are formed on a single thin wafer; substrate is called *monolithic* IC, which is very small in size. It requires microscope to see connection between components.

- The steps to fabricate IC chips is similar to the steps required to fabricate transistors, diodes etc. In IC chips, the fabrication of circuit elements such as transistors, diodes, capacitors etc. and their interconnections are done at the same time.

- IC has so many advantages such as extremely small size, small weight, low cost, low power consumption, high processing speed, easy replacement, etc. IC can function as amplifier, oscillator, timer, counter, computer memory etc.

2.7.2 Steps for IC Fabrication

- The manufacturing of a complete monolithic IC consists of number of following steps :

  1. Wafer processing
  2. Epitaxial growth
  3. Oxidation
  4. Photolithography
  5. Etching
  6. Diffusion
  7. Ion implantation
  8. Deposition
  9. Assembly and packing

- The steps include 8 to 20 patterned layers created into the substrate to form the complete monolithic IC.

- The electrically active regions are created due to this layering in and on the surface of wafer. Hundreds of integrated circuits can be made on a single thin silicon. Then it is cut into an individual IC chips.

1. Wafer Production :

- The steps used in the production of silicon wafer are as under :

  (i) Crystal growth and doping
  (ii) Ingot trimming and grounding
  (iii) Ingot slicing
  (iv) Wafer polishing and etching
  (v) Wafer cleaning

- A wafer is a round slice of semiconductor material such as silicon. It is more suitable for manufacturing due to its characteristics.

- Silicon wafer is the base or substrate for entire IC chip. First purified polycrystalline silicon is created from the sand by using *Czochrolaski crystal growth* process. Then it is heated in a furnace upto the temperature of 1420°C to melt the silicon.

- A small piece of solid silicon is dipped on the molten liquid. Then the solid silicon called a seed is slowly pulled from the melt. The liquid cools to form single crystal ingot.

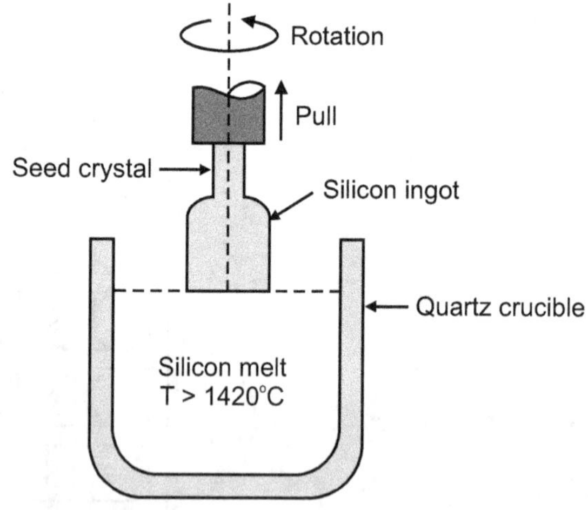

Fig. 2.23 : Czochrolaski crystal growth

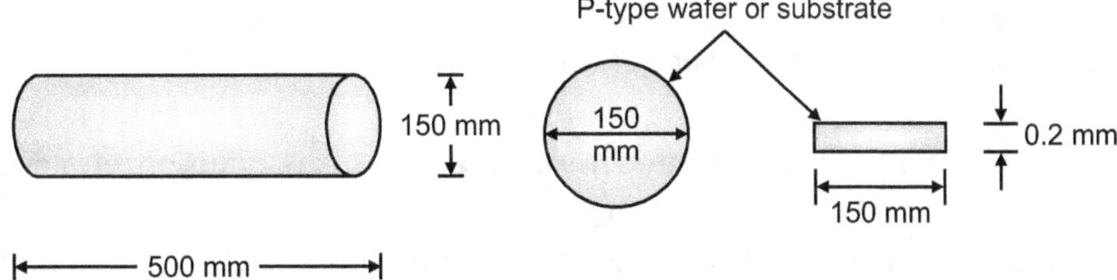

Fig. 2.24 : Wafer preparation

- A thin round wafer of silicon is cut using wafer slicer which is a precise cutting machine and each slice is having thickness of about 0.01 to 0.25 inches.

- When wafer is sliced, the surface will be damaged, it can be smoothened by polishing. After polishing the wafer, it must be thoroughly cleaned and dried. The wafers are cleaned using high purity low particle chemicals. The silicon wafers are exposed to ultra pure oxygen.

2. Epitaxial Growth :

- The word *epitaxy* is derived from Greek word *epi* meaning upon and the past tense of the word teinon meaning *arranged*. Thus, *epitaxy* means arranging atoms in a single crystal fashion upon a single crystal substrate, so that the resulting layer is an extension of the substrate crystal structure.

- Epitaxial growth is a chemical reaction used for growing of single silicon crystal upon original silicon substrate.

- A process is carried out in a epitaxial reactor as shown in Fig 2.25. The chamber is consisting of a long cylindrical quartz tube encircled by a long tube which is encircled by RF induction coil.

- The silicon is replaced on a rectangular rod called a *boat*, which is then inserted in the reaction chamber where the graphite is heated to a temperature of 1200°C.

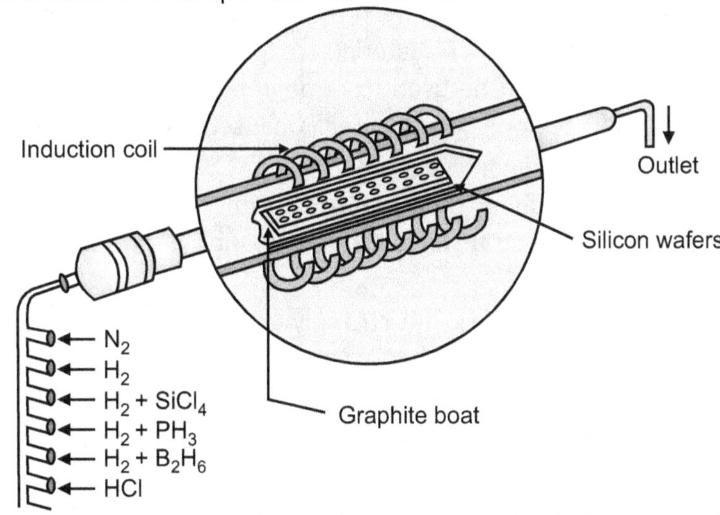

Fig. 2.25 : Epitaxial reactor for production of growth of Silicon epitaxial film

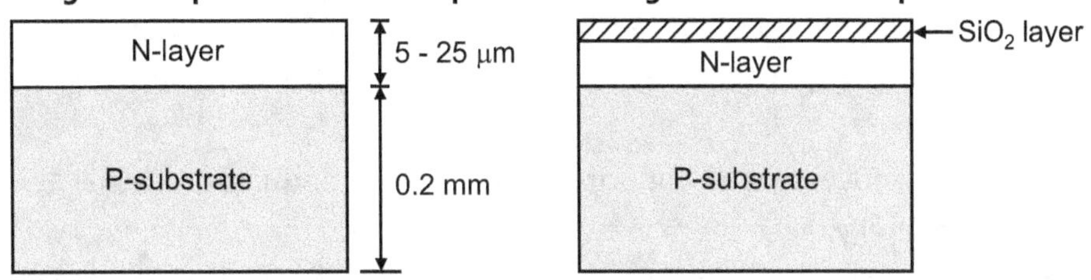

Fig. 2.26 : Epitaxial growth

- The various gases required for growth of desired epitaxial layer are introduced into the system through a control rod. Thus, it is possible to form almost an abrupt step in PN junction similar to an ordinary PN junction.

- An N-type silicon layer of about 5 to 25 μm is now grown on the P-type substance as N-type layer is grown by placing the silicon wafer in a special furnace called a **epitaxial reactor** at about 900°C to

1200°C, and the gases containing silicon gas serve as a source of silicon-phosphine gas, which is a source of phosphorus. Thus a uniform layer of silicon dioxide is formed on the surface of wafer.

3. Oxidation :

- The silicon-dioxide has the fundamental property of preventing the diffusion of almost all impurities through it. It serves two very important following properties :

    (i) Silicon-dioxide is an extremely hard protecting coating and is unaffected by almost all reagents except hydrofluoric acid. Thus it stands against any contamination.

    (ii) By selective etching of silicon-dioxide, the diffusion of impurities through carefully defined windows in silicon-dioxide can be accomplished to fabricate various components.

- Silicon wafers are stacked upto in a quartz *boat* and then inserted into quartz furnace tube. The silicon wafers are raised to a temperature in the range of 950°C to 1150°C and at the same time exposed to a gas containing O_2 or H_2O or both.

- Oxidation process is also called *thermal oxidation* because high temperature is maintained to grow the oxide layer.

4. Photolithography :

- To protect some area of wafer, when working on another area, a process called *photolithography* is used. The process of photolithography includes the masking with a photographic mask and photo etching.

- A photo resist film is applied on the wafer. The wafer is aligned to a mask using photo aligner. Then it is exposed to ultraviolet (UV) light through mask. Before that the wafer must be aligned with mask. Generally there are automatic tools for alignment purpose.

5. Etching :

- Etching is used to remove material selectively from the surface of wafer to create patterns. The pattern is defined by etching mask. The parts of material are protected by this etching mask. Either wet (chemical) or dry (physical) etching can be used to remove the unmasked material. To perform *etching* in all directions at same time, isotropic etching will be used. An isotropic etching is faster in one direction.

- Wet etching is isotropic, but the etching time control is difficult. Wet etching uses liquid solvents for removing materials. It is not suited to transfer pattern with submicron feature size. It does not damage the material.

- Dry etching uses gases to remove materials. It is strongly *anisotropic*. But it is less selective. It is suited to transfer pattern having small size.

- The remaining photoresist is finally removed using additional chemicals or plasma. Then the wafer is inspected to make sure that the image is transferred from mask to the top layer of wafer.

6. Doping :

- To alter the electrical character of silicon, atom with one less electron than silicon such as boron and atom with one electron greater then silicon such as phosphorus is introduced into the area.

- The P-type (boron) and N-type (phosphorus) are created to reflect their conducting characteristics.

7. Diffusion :

- Diffusion is defined as the movement of impurity atoms in semiconductor material at high temperature.

- The diffusion process generally takes place in two steps as under :

    (i) The high concentration of dopant atoms is brought in contact with the surface of a silicon wafer for a relatively short time. This is intended to bring the proper number of dopant atoms on the surface of the silicon wafer.

    (ii) The dopant surface is removed and under the influence of high temperature, the dopant atoms on the wafer surface penetrate the wafer, i.e. they are dissolved in the still-solid silicon crystal according to well established physical laws.

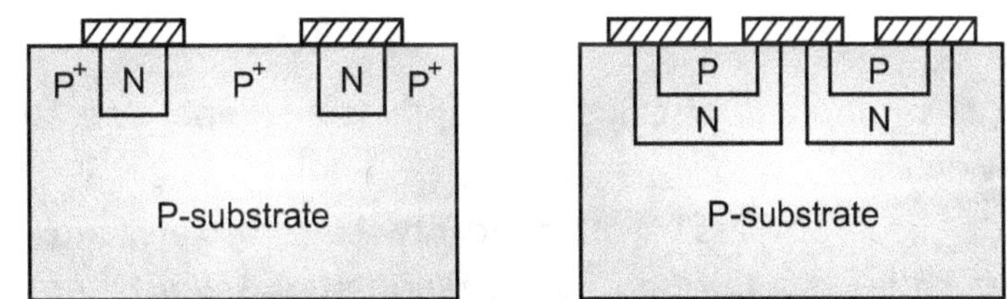

Fig. 2.27 : Diffusion of P-type impurities

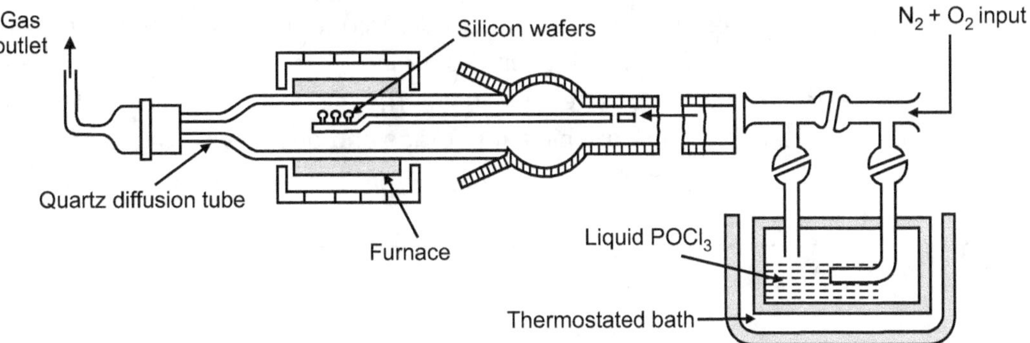

Fig. 2.28 : Apparatus for PCl₃ diffusion

- In *atomic diffusion* process, p and n regions are created by adding dopants into the silicon wafer. The wafers are placed in an oven, which is made up of quartz and it is surrounded with heating elements.

- Then the wafers are heated at a temperature of about 1500°C to 2200°F. The inert gas carries the dopant chemical. The dopant and gas is passed through the wafers and finally the dopant will get deposited on the wafer. This method can only be used for large areas. For small areas, it will be difficult and it may not be so accurate.

8. Ion Implantation :

- Ion Implantation is a method used for adding dopants in silicon wafers. In this method, dopant gas such as phosphine or boron trichloride will be ionized first. Then it provides a beam of high energy dopant ions to the specified regions of wafer. It will penetrate the wafer.

- The depth of penetration depends on the energy of the beam. By altering the beam energy, it is possible to control the depth of penetration of dopants into the wafer. The beam current and time of exposure is used to control the amount of dopant.

- This method is slower than atomic diffusion process. It does not require masking and this process is very precise.

9. Metallization (Deposition) :

- Metallization is used to create contact with silicon and to make interconnections on chip. A thin layer of aluminium is deposited over the whole wafer.

- Aluminium is selected because it is a good conductor, has good mechanical bond with silicon, forms low resistance contact and it can be applied and patterned with single deposition and etching process.

- The process such as masking, etching, and doping will be repeated for each successive layer until all integrated chips are completed.

- Between the components, silicon dioxide is used as an insulator. This process is called *chemical vapour deposition*. To make contact pads, aluminium is deposited. The fabrication includes more than three layers separated by dielectric layers.

- For electrical and physical isolation, a layer of solid dielectric is surrounded in each component which provide isolation.

- It is possible to fabricate PNP and NPN transistors in the same silicon substrate. To avoid damage and contamination of circuit, final dielectric layer (passivation) is deposited. After that, the individual IC will be tested for electric function and check the functionality of each chip on wafer. These chips are not passed in the test and will be rejected.

10. Assembly and Packaging :

- Each of the wafer contains hundreds of chips. These chips are separated and packaged by a method called *scribing* and cleaving.

- The wafer is similar to a piece of glass. A diamond saw cut the wafer into single chip. The diamond tipped tool is used to cut the lines through the rectangular grid which separates the individual chips. The chips that are failed in electrical test are discarded.

- Before packaging, remaining chips are observed under microscope. The good chip is then mounted into a package. Thin wire is connected using ultrasonic bonding. It is then encapsulated for protection. Before delivered to customer, the chip is tested again. There are three configurations available for packaging. They are metal can package, ceramic flat package and dual-in-line package.

- For military applications, the chip is assembled in ceramic packages. The complete integrated circuits are sealed in anti-static plastic bags.

2.8 MOS TECHNOLOGY

2.8.1 History

- The research work of *Frosch* and *Derick* on silicon dioxide to the metal oxide-semiconductor (MOS) transistor dominated semiconductor technology before 1963.

- After 1963, the combination of integrated circuit (IC) and the planer manufacturing process had led to MOS transistor as a potentially promising semiconductor technology.

- In the mid-1970s, the MOS transistor had established as a commercially successful and sustainable technology.

- *Fairchild* had put a lot of time and effort into studying the surfaces of bipolar transistors (BJTs), but the first sustained work on the MOS transistor as a 'potential' product come from *Frank Walnass* of Fairchild in August 1962.

- In December 1963, *Frank Walnass* explored the chemistry and physics of MOS structures, built MOS integrated circuits (ICs) and considered how various MOS phenomenon could be commercially exploited.

- The greatest technological contribution of *Frank Walnass* was the invention of complementary MOS (CMOS), which led to transistor circuits that consumed almost no power in standby operation.

- While the complexities of building CMOS circuits were so great in the 1960 is that most firms concentrated on making p-channel MOS circuits.

- After *Walnass's* MOS transistor work in early 1963, *Gordon Moore* began putting together a team to understand the MOS structure and the silicon-dioxide system in a systematic way. Upto this time, the problems of MOS stability were so great that the MOS transistor's characteristics might vary by over a hundred volts over time or with changes in temperature and operating conditions that made MOS transistors useless as a product.

- In October 1963, *Ed Snow* began a project assuming that different metals applied as a gate electrode over the silicon dioxide layer might show different levels of stability.

- *Ed Snow* used an electron beam evaporator to apply the various metals and he found more stability for aluminium structures. He also discovered that the key to stability was not the particular metal used, but the method of evaporating it.

- The research work of *Ed Snow* discovered that contaminated devices with sodium led to the highest drifts. So the minimized sodium in any metal used in the MOS production process.

- Between 1963 to 1967, *Deal, Grove* and *Ed Snow* along with *T. Sah* published over two dozen research papers related to the silicon-silicon dioxide interface.

- In 1968, Fairchild company had developed MOS memory chips, arithmetic-unit chips, calculator chips, and customizable logic chips.

- International Business Machine (IBM) developed and manufactured bipolar transistors (BJTs) for its large computer systems.

- By 1968, the companies such as American Microsystems or semiconductor operations of General Instrument focussed on MOS technology. IBM and Intel proved to be critical in establishing MOS technology.

- In 1968, *Edward Davis*, developed new computer memories with greater densities. In 1972, IBM introduced new computer systems using 1024 bit MOS memory chips.

- By 1974, MOS technology was fully established as a vital commercial technology. Intel had introduced its 4 kilobit memory chip and its second generation microprocessor, the popular 8080.

- In 1974, *Robert Dennard* clearly described in his research prefer the principle of device scaling to reduce the size about one-micron, which was good reason to think of MOS technology.

2.8.2 Introduction

- Metal-oxide-silicon (MOS) technologies have made a tremendous progress in the recent years. Presently MOS-technology is the dominating LSI technology.

- MOS-technology had developed MOS-transistors, inverters and the state of the art processing together with advanced transistor models for CAD applications. MOS-transistors have been scaled from long to short channels.

- CMOS-technology has become a competing VLSI-technology showing a large potential for the achievement of minimum power density, high speed and yet sufficient noise margins and of various compatibility integrated devices such as FETs and BJTs.

- The PMOS devices are based on the p-channel MOS transistors. Specifically, the PMOS channel is part of a n-type substrate lying between two heavily doped P^+ wells beneath the source and drain electrodes. General speaking, a PMOS transistor is only constructed in consert with an NMOS transistor.

- The NMOS technology and design processes provide an excellent background for other technologies. In particular, some familiarity with NMOS allows a relatively easy transition to CMOS technology and design.

- The techniques employed in NMOS technology for logic design are similar to GaAs technology.

2.8.3 MOS Structure

- CMOS knowledge is extensively used today to form circuits in various and diverse applications.

- Today's computers, CPUs and cell phones make use of CMOS due to numerous advantages. It deals with low power consumption, relatively high speed, high noise margins in both states, and will work over a wide range of source and input voltages, provided the source voltage is fixed.

- MOS structure comprises of the following three layers :

  1. Metal Gate Electrode
  2. Insulating Oxide Layer (SiO_2)
  3. P-type Semiconductor (Substrate)

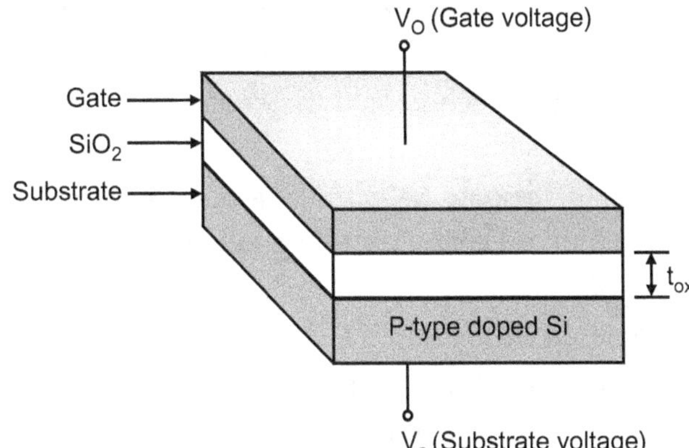

Fig. 2.29 : MOS structure

- MOS structure produces a capacitor, with gate and substrate as two plates and oxide layer as the dielectric material. The width of the dielectric material (SiO_2) is typically between 10 nm and 50 nm. Carrier concentration and delivery within the substrate can be operated by external voltage applied to gate and substrate terminal.

2.9 SILICON GATE TECHNOLOGY

2.9.1 History

- In 1966, *Dr. Brower* realized that if the gate electrode was defined first, it would be possible not only to minimize the parasitic capacitances between gate and source and drain, but it would also make them insensitive to misalignment.

- He proposed a method in which the aluminium gate electrode itself was used as a mask to define the source and drain regions of the MOS transistor.

- However, since aluminium could not withstand high temperature required for the conventional doping of the source and drain, *Dr. Brower* proposed to use ion implantation, a new doping technique.

- Dr. Brower's idea was not conceptually sound in practice it did not work, because it was impossible to adequately passivate the MOS transistors and repair the radiation damage done to the silicon crystal structure by the ion implantation, since these two operations would have required temperations in excess of the ones survivable by the aluminium gate. Thus, this invention proved a proof of principles but no commercial IC was ever produced with Brower's method. A more refractory material was needed.

- In 1967, *John C. Sarace* and collaborators at Bell Laboratories replaced the aluminium gate with an electrode made of vacuum-evaporated amorphous silicon and succeeded in building working self-aligned gate MOS transistors.

- However, the process as described was only a proof of principle suitable only for the fabrication of discrete transistors and not for ICs and was not any further by its investigators.

- In 1968, the MOS industry was prevalently using aluminium gate transistors with *High Threshold Voltage* (HTV) and desired to have a *Low Threshold Voltage* (LTV) MOS process in order to increase the speed and reduce the power dissipation of MOS ICs.

- Low threshold voltage transistors with aluminium gate demanded the use of silicon orientation, which however produced too flow a threshold voltage for the parasitic MOS transistors, which is created when aluminium over the field oxide would bridge two junctions.

- To increase the parasitic threshold voltage beyond the supply voltages, it was necessary to increase the N-type doping level in selected regions under the field oxide, and this was initially accomplished with the use of a so called *channel-stopper mask*, and later with ion implantation.

2.9.2 Introduction

- The Silicon Gate Technology (SGT) was first process technology used to fabricate commercial MOS ICs that was later widely adopted by the entire industry in the 1960s.

- The silicon gate technology was the world's first commercial MOS self-aligned-gate process technology. Before this technology, the control gate of the MOS transistor was made with aluminium instead of polycrystalline silicon.

- Aluminium-gate MOS transistors were three to four times slower, consumed twice as much silicon area, had higher leakage current and lower reliability compared with silicon-gate transistors.

- *Faggin* created the silicon gate technology in 1968. He also designed and built the world's first commercial IC using the silicon gate technology.

- A self-aligned gate is a transistor manufacturing feature, where by a refractory gate electrode region of MOSFET is used as a most for the doping of the source and drain regions.

- This technique ensures that the gate will slightly overlap the edges of the source and drain.

- The use of self-aligned gates is one of the many innovations that led to the large increase in computing power in the 1970s. Self-aligned gates are still used in most modern IC processes.

- The self-aligned gate is used to eliminate the need to align the gate electrode to the source and drain regions.

- One important feature of silicon gate technology was that the silicon gate was entirely buried under top quality thermal oxide which is one of the best insulators known, making it possible to create new device type, not feasible with conventional technology or with self-aligned gates made with other materials.

2.9.3 Manufacturing Process

- The process of using the gate oxide as a mask for the source and drain diffusion both simplify the process and greatly improve the yield.

- The following are the steps in creating a self-aligned gate: These steps were first created by *Fedrics Faggin* and used in the Silicon Gate Technology process developed at Fairchild Semiconductor in 1968 for the fabrication of the first commercial IC using it.

  1. Wells on the field oxide are etched, where the transistors are to be formed. Each well defines the source, drain and active gate regions of an MOS transistor.

  2. Using a *dry thermal oxidation process*, a thin layer of about 5 to 200 nm of gate oxide (SiO_2) is grown on the silicon wafer.

  3. Using a *chemical vapor deposition* (CVD) process, a layer of polysilicon is grown on top of the gate oxide.

  4. A layer of *photoresist* is applied on top of the polysilicon.

  5. A mask is placed on top of the photoresist and exposed to *ultraviolet (UV) light*. This breaks down the photoresist layer in areas, where the mask did not protect it.

  6. Photoresist is exposed with a specialized developer solution. This is intended to remove the photoresist that was broken down by the ultraviolet light.

  7. The polysilicon and gate oxide that is not covered by photoresist is etched away with a buffered ion etch process. This is usually an acid solution containing *hydrofluoric acid*.

  8. The rest of the photoresist is stripped from the silicon wafer. There is now a wafer with polysilicon over the gate oxide, and over the field oxide.

  9. The thin oxide is etched away exposing the source and drain regions of the transistor, except in the gate region which is protected by the polysilicon gate.

  10. Using a conventional doping process, or a process called ion implantation, the source, drain and the polysilicon are doped. The thin oxide under the silicon gate acts as a mask for the doping process. This step is what makes the gate self-aligning. The source and drain regions are automatically properly aligned with the gate.

  11. The wafer is *annealed* in a high temperature furnace at about greater than 800°C or 1,500°F. This diffuses the dopant further into the crystal structure to make the source and drain regions and results in the dopant diffusing slightly underneath the gate.

  12. The process continues with vapor deposition of silicon dioxide to protect the exposed areas, and with all the remaining steps to complete the process.

2.9.4 Benefits

- The benefits of the self-aligned silicon technology are as under :

  1. It is *faster* by a factor of 3 to 5 times over conventional MOS.

  2. Its *power dissipation* is reduced by a factor of 3 to 5 times over conventional MOS ICs.

  3. It is *cost effective* and its cost is reduced to about half that of a conventional MOS IC.

  4. It has *higher reliability* over conventional MOS ICs.

  5. It is possible to *fabricate new device* type with new functions.

2.10 CMOS TECHNOLOGY

2.10.1 Introduction

- A new type of MOSFET logic, Complementary metal–oxide–semiconductor (CMOS) was invented by *Chih-Tang Sah* and *Frank Walnass* at Fairchild Semiconductor in 1963.

- CMOS is a technology for constructing ICs, and is a form of MOSFET semiconductor. It refers to both, a particular style of digital circuitry design and the family of processes used to implement that circuitry on IC chips.

- CMOS circuits use a combination of PMOS and NMOS to implement logic gates and other digital circuits.

- Although CMOS logic can be implemented with discrete devices for demonstrations, commercial CMOS products are IC composed of upto billions of transistors of both types, on a rectangular piece of silicon of between 10 and 400 mm^2.

- CMOS always uses all *enhancement-mode* MOSFETs, in other words, a zero gate-to-source voltage turns the transistor off.

- Two important characteristics of CMOS devices are high *noise immunity* and low static *power consumption*.

- *CMOS* technology is used in *microprocessors, microcontrollers, static RAM* and other *digital logic circuits*. It is also used in for several analog circuits such as *image* (CMOS) *sensors, data converters*, and highly integrated *transceivers* for many types of communication. It is also widely used for *calculators* and *watches*.

2.10.2 NMOS

- An NMOS is a type of negative MOSFET (nMOSFET). An NMOS transistor is made up of n-type source and drain and a p-type substrate.

- When a voltage is applied to the gate, holes in the body (p-type substrate) are driven away from the gate. This allows forming an n-type channel carried by electrons from source and the drain and a drain current is carried by electrons from source to drain through an induced n-type channel.

- Logic gates and other devices implemented by using NMOS are said to have NMOS logic.

- There are three modes of operation in NMOS called the cut-off, triode and saturation.

- NMOS logic is easy to design and manufacture. But circuits with NMOS logic gates dissipate static power, when the circuit is idling, since DC current flows through the logic gate when the output is low.

2.10.3 PMOS

- A PMOS is a type of positive MOSFET (pMOSFET). A PMOS transistor is made up of p-type source and drain and an n-type substrate.

- When a positive voltage is applied between the source and the drain and the gate, a p-type channel is formed between the source and the drain with opposite polarities.

- A drain current is carried by holes from source to the drain through an induced p-type channel.

- A high voltage on the gate V_{GS} will cause a PMOS not to conduct, while a low voltage on the gate will cause it to conduct.

- Logic gates and other digital devices implemented using PMOS are said to have PMOS logic.

- PMOS technology is low cost and has a good noise immunity.

2.10.4 N-Channel MOSFET (NMOS) Transistor

- Consider the structure of the NMOS enhancement mode device as shown in Fig. 2.30.

- The two N-type regions are embedded in the P-type substrate. The N-type regions are the source and drain terminals.

- The region between the source and drain is the channel region, which is covered by a thin silicon dioxide layer.

- The gate is placed at the top of oxide layer.

- Because of the presence of the insulating silicon dioxide layer, the device is also known as isolated gate FET (IGFET). It is because of this layer, the MOSFETs have high input resistance.

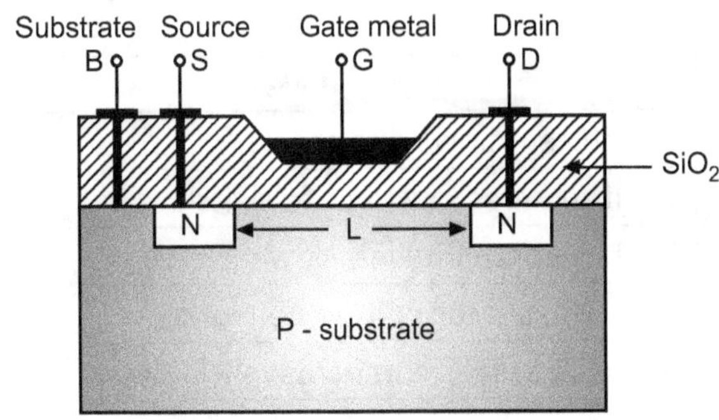

Fig. 2.30 : Enhancement mode NMOS

- A polysilicon gate is deposited on a layer of insulation over the region between source and drain.

- Fig. 2.30 shows a basic NMOS enhancement mode device in which the channel is not established and the device is in a non-conducting condition i.e. $V_D = V_S = V_{gs} = 0$.

- If this gate is connected to a suitable positive voltage with respect to the source, then the electric field established between the gate and the substrate gives rise to a charge inversion region in the substrate under the gate insulation and a conducting path or channel is formed between source and drain.

- The channel may also be established so that it is present under $V_{gs} = 0$ by implanting suitable impurities in the region between source and drain during manufacture and prior to depositing the insulation and the gate. This arrangement is shown in Fig. 2.30.

- Under these circumstances, the source and drain are connected by a conducting channel, but the channel may be closed by applying a suitable negative voltage to the gate.

- In both the cases, the variation of gate voltage allow control of any current flow between source and drain.

2.10.5 P-Channel MOSFET (PMOS) Transistor

- Consider the structure of PMOS enhancement mode device as shown in Fig. 2.31.

- The two P-type regions are embedded with N-type substrate. The P-type regions are the source and drain terminals.

- The region between source and drain is the channel region, which is covered by a thin SiO_2 layer. The gate is a placed at the top of oxide layer.

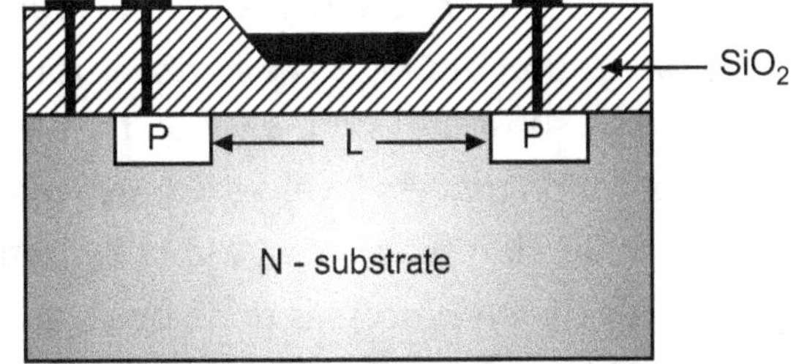

Fig. 2.31 : Enhancement mode PMOS

- Because of the presence of insulating SiO_2 layer, the device is also known as insulated gate FET (IGFET). It is because of this layer, the MOSFETs have a high input resistance.

- Fig. 2.31 shows a basic PMOS enhancement mode device in which the channel is not established and the device is in a non-conducting condition, i.e. $V_D = V_S = V_{GS} = 0$.

- If the gate is connected to a suitable negative voltage with respect to the source, then this gives rise to the formation of a channel (P-type) between the source and drain and current may then flow if the drain is made negative with respect to the source.

- In this case, the current is carried by holes as opposed to electrons (as is the case of NMOS devices). In consequence, PMOS transistors are inherently slower than NMOS, since hole mobility is less by a factor of approximately 2.5, than electron mobility.

- However, bearing these differences in mind, the discussions of NMOS transistors which follow relate equality well to PMOS transistors.

2.11 COMPARISON OF NMOS AND PMOS

Table 2.17

PMOS	NMOS
1. Current flows due to fets.	1. Current flows due to electrons.
2. It requires low logic to switch.	2. It requires high logic to switch.
3. It is comparatively slower.	3. It is faster.
4. It is used for pull-up network.	4. It is used for pull-down network.
5. Its IC area is comparatively larger.	5. Its IC area is smaller.

2.12 CMOS FABRICATION PROCESSES

2.12.1 Introduction

- In early 1960's the semiconductor manufacturing process was initiated from Texas and in 1963, CMOS was patented by *Frank Wanlass*.

- ICs are manufactured by utilizing the semiconductor fabrication process. These ICs are major components of every electrical and electronic devices, which we use in our daily life.

- Many complex and simple electronic circuits are being designed on a wafer made of semiconductor compounds and mostly silicon by using different fabrication steps.

- The CMOS technology is used in developing the microprocessors, microcontrollers, digital logic circuits and many other ICs.

- The CMOS technology facilitates low-power dissipation and high-packing density to build digital circuitry.

2.12.2 Fabrication Process Steps

- The CMOS can be fabricated using different processes such as :
  1. N-well process   2. P-well process   3. Twin tub process
- A P-well has to be created on an N-substrate or N-well has to be created on a P-substrate.

- Here fabrication of CMOS is described using the P-substrate, in which the NMOS transistor is fabricated on a P-type substrate and the PMOS transistor is fabricated N-well.

- The fabrication process involves twenty steps, which are as follows :
  1. Primarily, start the process with a P-substrate.
  2. The oxidation process is done by using high-purity oxygen and hydrogen, which are exposed in an oxidation furnace approximately at 1000°C.
  3. A light-sensitive polymer that softens, whenever exposed to light is called as photoresist layer.
  4. The photoresist is exposed to ultraviolet rays through the N-well mask.
  5. A part of the photoresist layer is removed by treating the wafer with the basic or acidic solution.
  6. The silicon-dioxide (SiO_2) layer is removed through the open area made by the removal of photoresist using hydrofluoric acid.
  7. The entire photoresist layer is stripped off.
  8. By using ion implantation or diffusion process, N-well is formed.
  9. Using hydrofluoric acid, the remaining SiO_2 layer is removed.
  10. Chemical Vapor Deposition (CVD) process is used to deposit a very thin layer of gate oxide.
  11. Except the two small regions required for forming the gates of NMOS and PMOS, the remaining layer is stripped off.
  12. Next, an oxidation layer is formed on this layer with two small regions for the formation of the gate terminals of NMOS and PMOS.

13. By using the masking process small gaps are made for the purpose of N-diffusion. The n-type (n^+) dopants are diffused or ion implanted, and the three n^+ are formed for the formation of the terminals of NMOS.

14. The remaining oxidation layer is stripped off.

15. Similar to the above N-diffusion process, the P-diffusion regions are diffused to form the terminals of PMOS.

16. A thick-field oxide is formed in all regions except the terminals of PMOS and NMOS.

17. Aluminium is sputtered on the whole wafer.

18. The excess metal is removed from the wafer layer.

19. The terminals of the PMOS and NMOS are made from respective gaps.

20. Assigning the names of the terminals of the NMOS and PMOS.

2.13 CMOS FABRICATION TECHNOLOGIES

2.13.1 Types of Technologies

- CMOS fabrication can be accomplished using either of the four technologies:

  1. Twin well/Tub Technology
  2. Silicon On Insulator Technology
  3. N-well Technology
  4. P-well Technology

2.13.2 Twin-well/Tub Technology

- The logical extension of the p-well and n-well technologies is the Twin-well technology. Thus, we can optimize NMOS and PMOS transistors separately. This means that transistor parameters such as threshold voltage, body effect and the channel transconductance of both types of transistors can be tuned independenly.

- n^+ or p^+ substrate with a lightly doped epitaxial layer on top forms the starting material for this technology. The n-well and p-well are formed on this epitaxial layer which forms the actual substrate.

- The dopant concentrations can be carefully optimized to produce the desired device characteristics because two independent doping steps are performed to create the well regions.

- The conventional n-well CMOS process suffers from among other effects, the problem of unbalanced drain parasitics since the doping density of the well region typically being about one order of magnitude higher than the substrate. This problem is absent in the twin-tub process.

2.13.3 Silicon On Insulator (SOI) Process

- To improve process characteristics such as *speed* and *latch-up susceptibility*, technologists have sought to use an insulating substrate instead of silicon as the substrate material.

- Completely isolated NMOS and PMOS transistors can be created virtually side by side on an insulating substrate eg. sapphire by using the SOI CMOS technology.

- This technology offers advantages in the form of higher integration density because of the absence of well regions, complete avoidance of the latch-up problem, and lower parasitic capacitance compared to the conventional n-well or twin-tub CMOS processes.

- But this technology comes with the disadvantage of higher cost than the standard n-well CMOS process. Yet the improvements of device performance and the absence of latch-up problems can justify its use, especially in deep submicron devices.

2.13.4 N-well Technology

- N-well CMOS fabrication technology is a well established technology, which requires both n-channel and p-channel transistors be built on the same chip substrate.

- To accomodate this, special regions are created with a semiconductor type opposite to the substrate type. The regions thus formed are called *wells* or *tubs*.

- In an n-type substrate, we can create a p-well or alternatively, an n-well is created in a p-type substrate. We present here a simple n-well CMOS fabrication technology, in which the NMOS transistor is created in the p-type substrate, and the PMOS in the n-well, which is built-in into the p-type substrate.
- Fabrication started with p-well technology but now it has been completely shifted to n-well technology. The main reason for this is that, "n-well sheet resistance can be made lower than that of p-well sheet", because electrons are more mobile than holes.

Process Sequences :

- The simplified process sequence for the fabrication of CMOS ICs on a p-type silicon substrate is as follows:
    1. N-well regions are created for PMOS transistors, by impurity implantation into the substrate.
    2. This is followed by the growth of a thick oxide in the regions surround the NMOS and PMOS active regions.
    3. The thin gate oxide is subsequently grown on the surface through thermal oxidation.
    4. After this n^+ and p^+ regions (source, drain and channel-step implants) are created.
    5. The metallization step (creation of metal interconnects) forms the final step in this process.

Advantages :

- The advantages of n-well process are as follows :
    1. N-well CMOS is superior to P-well because of the lower substrate bias effect on transistor threshold voltage.
    2. It has lower parasitic capacitances associated with source and drain regions.
    3. Latch-up problems can be considerably reduced by using a how resistivity epitaxial p-type substrate.
    4. N-well process degrades the performance of poorly performing p-type transistor.

2.13.5 P-well Process

- Among all the fabrication processes of the CMOS, N-well process is mostly used for the fabrication of the CMOS. P-well process is almost similar to the N-well.
- But the only difference in p-well process is that it consists of a main N-substrate and, thus, P-well itself acts as a substrate for N-devices.
- The basic processing steps of the P-well process are of the same nature used for NMOS. The structure consists of an n-type substrate in which P-device may be formed by suitable masking and diffusion, and in order to accommodate n-type devices, a deep p-well is diffused into the n-type substrate.
- The diffusion must be carried out with special care since the p-well doping concentration and depth will affect the threshold voltages as well as the breakdown voltages of n-transistors.
- To achieve low threshold voltages of 0.6 to 1.0 V, we need either deep well diffusion or high well resistivity. Deep well requires larger spacing between the n and p-type transistors and wires because of lateral diffusion resulting in larger chip area.
- The P-well acts as a substrate for n-devices within the parent n-substrate and provided that voltage polarity restrictions are observed, the two areas are electrically isolated.

Processing Steps :

- Since there are now in effect two substrate connections, V_{DD} and V_{GS} are required. The processing steps for P-well fabrication are as follows :
    1. Mask 1 defines the areas in which the deep p-well diffusions are to take place.
    2. Mask 2 defines the twin oxide regions. The areas where the thick oxide is to be stripped and thin oxide grown to accommodate p and n transistors and diffusion wires.
    3. Mask 3 is used to pattern the polysilicon layer, which is deposited after the thin oxide.
    4. A p-plus mask is used to define all areas, where p-diffusion is to take place.
    5. Mask 5 is performed using the negative form these areas where n-type diffusion to take place.

6. Mask 6 defines the contact cuts.

7. The metal layer pattern is defined by thin mask 7.

8. An overall passivation over glass layer is now applied and mask 8 is needed to define the openings for access to bonding pads.

Practice Questions

1. List BJT and CMOS parameters.

2. Explain working of CMOS inverter with its circuit diagram.

3. Draw circuit diagram of following logic gates using CMOS inverters :

    (a) OR, (b) AND, (c) NOR, (d) NAND

4. Explain transmission gate with its diagram.

5. Explain MOS transistor switches with their circuit diagram.

6. Draw CMOS inverter voltage transfer characteristic and explain it.

7. Explain resistor and capacitor layout.

8. Compare CMOS and BJT technologies.

9. List IC fabrication processes.

10. Component PMOS and NMOS transistors.

11. Explain working of PMOS and NMOS transistors with their structure.

12. What are advantages of silicon gate technology?

13. List 20 steps of CMOS fabrication process.

14. Compare BJT and CMOS parameters.

15. List IC fabrication processes.

16. Describe the following processes :

    (a) wafer processing, (b) oxidation, (c) epitaxy, (d) deposition, (e) diffusion, and (f) ion implantation.

17. Describe silicon gate process of CMOS fabrication.

18. Describe how integrated resistors and capacitors are fabricated.

19. Compare PMOS and NMOS transistors.

20. Explain working of N-channel MOSFET in depletion and enhancement mode with its structures.

21. Explain working of P-channel MOSFET in depletion and enhancement mode with its structures.

22. What is threshold voltage ?

23. Write PMOS technology and draw MOS structure.

24. Explain the working of PMOS and NMOS transistors with its structure.

25. What are the benefits of silicon gate technology ?

26. What are CMOS transistors ?

27. Derive a MOS device design equation.

28. List 20 steps of CMOS fabrication processes.

29. List four CMOS fabrication technologies.

30. Describe following CMOS fabrication technologies :

    (a) Twin-tub, (b) Silicon on insulator, (c) N-well, (d) P-well.

31. List advantages of N-well technology.

32. What is CMOS logic gate ?

33. Explain working of 2-input CMOS logic gates :

    (a) Inverter, (b) OR, (c) AND, (d) NAND, (e) NOR with logic diagram.

34. Explain working of following 3-input CMOS logic gates with logic diagram :

    (a) OR, (b) AND, (c) NAND, (d) NOR.

35. What are advantages and disadvantages of CMOS ICs ?

36. List applications of CMOS ICs.

37. Explain working of CMOS transmission gate with its logic circuit and truth table.

38. Design following Boolean functions using CMOS logic gates :

    (a) $Y = A + BC$, (b) $Y = \overline{A} \, (\overline{B} + \overline{C})$, (c) $Y = AB + \overline{C}$, (d) $Y = (\overline{A} + \overline{B}) \, C$, (e) $Y = \overline{AB + CD}$

INTRODUCTION TO VHDL

3.1 Introduction to HDL : History of VHDL, Pro's and Con's of VHDL.

3.2 VHDL Flow Elements : Entity, Architecture, Configuration, Package, Library only definitions.

3.3 Data types, Operators, Operations.

3.4 Signal, Constant and Variables (Syntax and use).

3.5 VHDL modeling : Data flow, Behavioral, Structural.

(3a) Describe hardware description language, its components and programming syntax.

(3b) Describe the given VHDL flow elements.

(3c) Describe the use of given data type declaration in VHDL.

(3d) Describe the given type of VHDL modeling.

3.1 VERY LARGE SCALE INTEGRATION (VLSI)

3.1.1 History

- With the invention of transistors at Bell Laboratories in U.S.A in 1947, the field of electronics shifted from vacuum tubes to solid-state devices.

- Electrical engineers of the 1950s saw the possibilities of constructing far more advanced circuits. However, as the complexity of circuits grew, problems arose. One problem was the size of the circuit.

- A complex circuit like a computer was dependent on speed. If the components were large, the wires interconnecting them must be long. The electric signals took time to go through the circuit, thus slowing the computer.

- The invention of the integrated circuit (IC) by **Jack Kilbey** and **Robert Noyce** in 1959 solved this problem by making all the components and the chip out of the same block (monolith) of semiconductor material.

- The circuit could be made smaller, and the manufacturing process could be automated. This led to the idea of integrating all components on a single-crystal of silicon wafer, which led to *small-scale integration* (SSI) in the early 1960s, and then *medium-scale integration* (MSI) in the late 1960s, and then *large-scale integration* (LSI) as well as *very large scale integration* (VLSI) in the 1970s and 1980s respectively, with tens of thousands of transistors on a single chip (later hundreds of thousands, then millions, and now billions).

- Current technology has moved far past this mark and today's **microprocessors** have many millions of logic gates and billions of individual transistors.

- At one time, there was an effort to name and calibrate various levels of *large-scale integration* (LSI) above VLSI. Terms like *ultra-large-scale integration* (ULSI) were used. But the huge number of gates and transistors available on common devices has rendered such fine distinctions meet. Terms suggesting greater than VLSI levels of integration are no longer in widespread use.

- In 2008, billion-transistor processors became commercially available. This became more common place as semiconductor fabrication advanced from then-current generation of 65 nm processes.

- Current designs, unlike the earliest devices, use extensive **design automation** and automated **logic synthesis** to lay out the transistors, enabling higher levels of complexity in the resulting logic functionality.
- Certain high-performance logic blocks like the SRAM (Static Random-Access Memory) cell, are still designed by hand to ensure the highest efficiency.

3.1.2 Structured Design

- Structured VLSI design is a modular methodology originated by **Carver Mead** and **Lynn Conway** for saving microchip area by minimizing the interconnect fabric area.
- This is obtained by repetitive arrangement of rectangular macro blocks which can be interconnected using wiring by **abutment**. In complex designs this structuring may be achieved by hierarchical nesting.
- Structured VLSI design had been popular in the early 1980s, but lost its popularity later because of the advent of placement and routing tools wasting a lot of area by routing, which is tolerated because of the progress of **Moore's Law**.
- When introducing the hardware description language, **KARL** in the mid 1970s, **Reiner Hartenstein** coined the term *"structured VLSI design"*, echoing **Edsger Dijkstra's** structured programming approach by procedure nesting to avoid chaotic spaghetti-structured program.

3.1.3 Definition

- VLSI stands for Very Large Scale Integration. **VLSI is the process of creating an integrated circuit (IC) by combining millions of transistors or devices into a single chip.** The microprocessor is a VLSI device.
- Before the introduction of VLSI technology, most ICs had a limited set of functions, they could perform.
- ***VLSI is the technology used to create small microchips, which are capable of performing various operations*** e.g., multiplexers, encoders, flipflops.

3.1.4 Advantages

1. Reduces the size of circuits
2. Reduces cost of devices
3. Increases operating speed of circuits
4. Less power consumption
5. High reliability
6. Occupies less area

3.1.5 Applications

1. It is used in a microprocessor of a personal computer or work station.
2. It is used as a chip in a graphic card, digital camera or camcorder.
3. It is used as a chip in a cell phone as a portable computing device.
4. It is used as embedded processor in an automobile.
5. It is used in the applications of artificial intelligence (AI) such as an expert systems, which work as design assistants.

3.2 HARDWARE DESCRIPTION LANGUAGE (HDL)

3.2.1 History

- The first Hardware Description Languages (HDL) appeared in the late 1960s, looking like more traditional languages.
- The first lasting effect was described in the text Computer Structures by **C. Gordon Bell** and **Allen Newell's** in 1971. This text introduced the concept of **register transfer level**, first used in the ISP language to describe the behavior of the **Digital Equipment Corporation** (DEC) PDP-8.
- The language became more widespread with the introduction of DEC PDP-16 RT-Level Modules (RTMs) and a book describing their use.
- University of Kaiserslautern produced a language called KARL (KAiserslautern Register Transfer Language), which included design calculus language features supporting VLSI chip floor planning and structured hardware design.
- In the mid-1980s, a VLSI design framework was implemented around KARL and ABL (A Block diagram language) by an international consortium funded by the Commission of the European Union.

- As design shifted to VLSI, the first modern HDL, **Verilog**, was introduced by **Gateway Design Automation** in 1985.
- In 1987, a request from the U. S. Department of Defence led to the development of VHDL (VHSIC Hardware Description Language).
- Initially, Verilog and VHDL were used to document and simulate circuit designs already captured and described in another form such as *schematic* files. HDL simulation enabled engineers to work at a higher level of abstraction than simulation at the schematic level, and thus increased design capacity from hundreds of transistors to thousands.
- Within a few years, VHDL and Verilog emerged as the dominant HDLs in the electronics industry, while older and less capable HDLs gradually disappeared from use.

3.2.2 Introduction

- A hardware description language (HDL) enables a precise, formal description of an electronic circuit that allows for the automated *analysis* and *simulation* of an electronic circuit.
- It also allows for the *synthesis* of a HDL description into a *netlist*, which can then be *placed and routed* to produce the *set of masks* used to create an *integrated circuit (IC)*.
- A HDL looks much like a *programming language* such as C; it is a textual description consisting of expressions, statements and control structures.
- HDLs form an integral part of *Electronic Design Automation* (EDA) systems, especially for complex circuits, such as *application-specific integrated circuits*, microprocessors, and *programmable logic devices*.
- HDLs are used to write executable specifications for hardware. HDLs are more precisely classified as *specification languages* or *modeling languages*.

3.2.3 Need

- *Moore's Law* in the year 1970 has brought a drastic change in the field of IC technology. This change has made the developers to bring out complex digital and electronic circuit. But the problem was the absence of a better programming language allowing hardware and software codesign.
- Complex digital circuit designs require more time for development, synthesis, simulation and debugging. The arrival of HDLs has helped to solve this problem by allowing each module to be worked by a separate team.
- All the goals like power, throughput, latency (delay), test coverage, functionality and area consumption required for a design can be known by using HDL.
- As a result, the designer can make the necessary engineering tradeoffs and can develop the design in a better and efficient way. Simple syntax, expressions, statements, concurrent and sequential programming is also necessary while describing the electronics circuits. All these features can be obtained by using a hardware description language. Now while comparing HDL and C languages, the major difference is that HDL provides the timing information of a design. Therefore, there is a need of Hardware Description Language (HDL)

3.2.4 Definition

- **A Hardware Description Language (HDL) is a specialized computer language used to describe the structure and behavior of electronic circuits and most commonly, digital logic circuits.**

3.2.5 HDL Languages

- The most common HDL languages are as under :
  1. Verilog
  2. VHDL
  3. System C
- The HDL will allow fast design and better verification. In most of the industries, Verilog and VHDL are common.

- Verilog standardized as IEEE 1364 is used for designing all types of circuits.
- VHDL (Very High Speed Integrated Circuit Hardware Description Language) is standardized by IEEE 1964.

3.2.6 Benefits

1. Fast design
2. Better verification
3. Reduced design cost
4. Reduced design time
5. Reduced design errors
6. Improved efficiency

3.2.7 Advantages

1. Verify design functionality easily in the design written as an HDL description.
2. Design simulation at higher level before implementation at gate level.
3. Allows test architecture and design decision.
4. Reusability of resources and design.
5. Increased flexibility to design changes.
6. Better and carrier design audition and verification.
7. Reduced non-recurring engineering costs.
8. Provides timing information.
9. Allows architectural design to be described in gate level and register transfer level.

3.2.8 Applications

1. It can be used to support the design and simulation of digital systems.
2. It is used to write executable rectifications for hardware.
3. It is most widely used in industry for digital circuit design.
4. It is used to describe the architecture and behavior of describe electrons circuits.

3.3 VHDL

3.3.1 History

- VHDL was originally developed at the behest of the U.S Department of Defense in order to document the behavior of the ASICs.
- The initial version of VHDL, designed in 1987 to IEEE standard IEEE 1076-1987, included a wide range of data types, including numerical, logical, character and time, plus arrays of bit called ***bit_vector*** and of character called ***string***.
- This IEEE standard 1164, which defined the nine-value, Logic types such as scalar std_logic and its vector version.
- The updated IEEE-1076 standard was developed in 1993, which made the syntax more consistent, allowed more flexibility in naming, extended the character type to allow ISO-8859-1 printable characters, added the operator, etc.
- Minor changes in the standard were made in 2000 and 2002 to add the idea of protected types similar to the concept of class in C++ and removed some restrictions from port mapping rules.
- In addition to IEEE standard 1164, several child standards were introduced to extend functionality of the language. IEEE 1076.2 added better handling of real and complex data types. IEEE standard 1076.3 introduced signed and unsigned types to facilitate arithmetical operations on vectors. IEEE standard 1076.1 known as VHDL-AMS.

3.3.2 Introduction

- To reduce the digital design in small space, various electronic hardware were used. Now instead of using hardware for design, software languages like verilog and VHE IC were used. The languages VHSIC is known as VHDL.

- The software languages VHDL was first introduced in 1981 for the Department of Defense (DoD) in U.S.A. under VHSIC program.
- VHDL is the VHSIC Hardware Description Language. VHSIC is an abbreviation for **V**ery **H**igh **S**peed **I**ntegrated **C**ircuit.
- VHDL can describe the behaviour and structure of electronic systems, but is particularly suited as a language to describe the structure and behavior of digital electronic hardware designs, such as ASICs and FPGAs as well as conventional digital circuits.
- VHDL is not an information model, a database schema, a simulator, a toolset or a methodology! However, a methodology and a toolset are essential for the effective use of VHDL. Simulation and synthesis are the two main kinds of tools, which operate on the VHDL language.
- VHDL does not constrain the user to one style of description. It allows designs to be described using any methodology such as top down, bottom up or middle out! VHDL can be used to describe hardware at the gate level or in a more abstract way.

3.3.3 Definition

- **VHDL is a programming language used to model a digital system by dataflow, behavioral and structural style of modeling.**
- *VHDL is a Hardware Description Language (HDL) used for modeling digital systems made of interconnection of components.*
- VHDL is an industry standard language used to describe the hardware from abstract to the concrete level and has become the communication medium of design for specifying input and output from various design tools.

3.3.4 Need

- Prior to VHDL, there were a number of proprietary languages for hardware description. It was created as an open standard as part of the government funded VHISC program so that designs were portable between design environments.
- EDIE was another effort in portability from the same, but it dealt mostly with structural data rather than behavior. Using VHDL, we can quickly describe and synthesize circuit of 20 thousands or more gates on a single chip. VDHL provides the capabilities i.e. advantages. Therefore, there is a need of VHDL for digital design.
- Therefore, there is a need of VHDL.

3.3.5 Features

1. It basically works in concurrent form and execute of the same time in parallel as in hardware.
2. It may also execute sine at a time in sequence as in any conventional language.
3. It produces net in EDIF (Electronic Design Interchange Forecast) after synthesizing VHDL code and it is need to exchange graphical and electronic data between EDH tools from various vendors.
4. It supports both synchronous and asynchronous timing models.
5. It servers as a communication medium between different CAD and CAE tools.
6. It supports language hierarchy, i.e. digital systems modeling as a set of interconnected components.
7. It supports various hardware technologies.
8. It changes medium between chip vendors and CAD tool mers.
9. It supports flexible design methodologies such as top-down, bottom-up or mixed.
10. It makes large scale design easier.
11. Test benches can be written in the same languages and circuit can be verified by simulation before synthesis.
12. Propagation delays minimum and maximum delays set-up and hold, time ting constrains can be described naturally.

13. The analyzed designed code can be inserted into a design library, which stress analyzed VHDL, descriptions.

14. It supports three types of modeling styles. It has powerful constraints.

15. The digital design may be decamped hierarchically. Each design element has a well defined interface.

3.3.6 Pro's and Con's

1. VHDL can be used as a communication medium between different CAD and CAE tools, between chip vendors and CAD tool users.

2. VHDL supports hierarchy, that is, a digital system can be modeled as a set of interconnected components, each component, in turn, can be modeled as a set of interconnected subcomponents.

3. It supports both synchronous and asynchronous timing models.

4. It is publicly available, human-readable, machine-readable and above all it is not proprietary.

5. As it is IEEE and ANSI standard, therefore, models described using this language are portable.

6. It supports three basic different description styles – structural, dataflow and behavioral.

7. Arbitrarily large designs can be modeled using the VHDL and there are no limitations on the size of a design.

8. Test benches can be written using the same language to test other VHDL models.

9. Nominal propagation delays, min-max delays, set up and hold timing, timing constraints and spike detection can also be described very naturally in this language.

10. It is not technology specific, but is capable of supporting technology specific features.

11. VHDL is not a *case sensitive* means it differentiate between uppercase and lowercase.

3.3.7 Capabilities

1. It can be used as a communication medium between different chip vendors and CAD tool users.

2. It can also used as a communication medium between different CAD and CAE tools.

3. It supports hierarchy, that is, a digital system can be modeled as a set of interconnected components; each component, in turn, can be modeled as a set of interconnected subcomponents.

4. It supports flexible design methodologies such as top-down, bottom-up, or mixed.

5. It is not technology-specific, but is capable of supporting technology-specific features.

6. It can also support various hardware technologies. By being technology independent, the same behavior model can be synthesized into different vendor libraries.

7. It supports both synchronous and asynchronous timing models.

8. It is publicly available, human readable, machine readable, and above all, it is not proprietary.

9. It supports a wide range of abstraction levels ranging from abstract behavioral descriptions to very precise gate-level descriptions.

10. It allows a design to be captured at a mixed level using a single coherent language.

11. Arbitrarily large designs can be modeled using the language and there are no limitations that are imposed by the language on the size of a design.

12. It does not support modeling at or below the transistor level.

13. The language has elements that make large scale design modeling easier, for example, components, functions, procedures, and packages.

14. There is no need to learn a different language for simulation control.

15. Test benches can be written using the same language to test other VHDL models.

16. Nominal propagation delays, min-max delays, setup and hold timing, timing constraints, and spike detection can all be described very naturally in this language.

17. Generics and attributes are also useful in describing parameterized designs.

18. A model cannot only describe the functionality of a design, but can also contain information about the design itself in terms of user-defined attributes, for example, total area and speed.

19. A common language can be used to describe library components from different vendors.

20. Tools that understand VHDL models will have no difficulty in reading models from a variety of vendors since the language is a standard.

21. Models written in this language can be verified by simulation, since precise simulation semantics are defined for each language construct.

22. Behavioral models that conform to a certain synthesis description style are capable of being synthesized to gate-level descriptions.

23. The capability of defining new data types provides the power to describe and simulate a new design technique at a very high level of abstraction without any concern about the implementation details.

3.3.8 Advantages

1. It is standard based, and transportable to other platforms.
2. Its carrier offer design modifications.
3. It is easiest for data entry or truth table results of Boolean entry.
4. Its structure is very similar to other programming languages.
5. It is easier with an intermediate and also need digital electronic background.
6. With experience it is less time consuming for more compiler circuits.
7. It is easy to translate flow chart into a functional circuit design.
8. It has greater control of practical issues such as time delay, latency, design changes etc.
9. With a template or example it can be quite easy to use.
10. It has powerful language construct and flexible.
11. It provides device independent or technology independent design.
12. It makes possible to change development tools.
13. It has lower cost and quick time to market.

3.3.9 Disadvantages

1. Synthesis results of VHDL may vary from one tool to another.
2. VHDL compilers do not produce optimal implementation. The optimal solution depends on the design objectives.
3. It has poor implementation due to the results of inefficient code.
4. The inefficient VHDL code results in repetitive unneeded and non-optimal logic.
5. For analog electronics, the VHDL is not yet standardized.
6. The descriptions of these types are neither high-level nor device independent.

3.3.10 Applications

1. To provide human and machine readable documentation.
2. To provide a structure reflecting hardware design and hierarchy.
3. To provide methods to handle complexity by partitioning the design.
4. To provide a concurrent a current method varying hardware interaction.
5. To provide ability to generate test vectors, multiple testbench strategies and method to write self checking code.
6. To provide high-level constructs that can be translated to Boolean equations, then translate them to gate as well as to provide constructs that can be optimized.

7. For development of Application Specific Integrated Circuits (ASICs).

8. Tools for automatic terms formation of VHDL code into a gate level net list, which is called *synthesis*. For design of low complexity Programmable Logic Devices (PLDs).

9. To provide standard solutions, e.g. for micro controllers, error correction decodes, etc. or behavioral models of microprocessor and RAM decries.

10. As a communication medium between different CAD and CAE tools, and also between chip vendors and CAD tool us.

11. To model a digital system as a set of interconnected components and to model each compress a set of interconnected subcomponents.

12. To write code descriptions of complex control logic.

13. To write test benches to test other VHDL models.

14. To describe library components from different vendors.

3.4 COMPARISON OF VHDL AND VERILOG

Table 3.1

VHDL	Verilog
1. It is somewhat difficult and complex.	1. It is simpler.
2. It has a concept of library.	2. It has no concept of library.
3. It allows concurrent procedure calls.	3. It does not allow task calls
4. It results slower simulation.	4. It results tools simulation.
5. It has entity deal a ration.	5. It has modulated deal a ration.
6. It has a concept of package.	6. It has no concept of package.

3.5 VHDL BASIC ELEMENTS

3.5.1 Introduction

- VHDL is a strongly typed language, which assists designers to catch, errors early in the development cycle. The compiler's analyzer is very exact and displays the errors for not using the correct data representation.

- In many aspects, VHDL is like a programming language. It draws its facilities from the familiar programming languages.

- VHDL is generally a case sensitive, which means that *lower case* and *upper case* letters are not distinguished. This can be exploited to define own rules for formatting the VHDL source code. VHDL keyword could, for example, be written in lower case letters and self defined identifiers in upper case letters.

- Statements are terminated in VHDL with a semicolon that means as many times breaks or other constructs as wanted can be inserted or left out. Only the semicolons are considered by the VHDL compiler.

- List is normally repeated by commas, signal assignments are noted with the composite assignment operator '< ='.

3.5.2 Basic Language Elements

- The types of language elements of VHDL are as under :

  1. Identifier.　　　　　　　　　　2. Object types.

  3. Data types.　　　　　　　　　　4. Operators.

3.6 IDENTIFIERS

3.6.1 Introduction

- An identifier in VHDL is a user defined name for constant, variable, signal, function, entity, subprogram etc. Thus it is used to name items for a VHDL model.
- An identifier can be any length in other words, as many characters as desired.
- An identifier is a *case insensitive*, meaning that there is no difference between lower case letters and upper case letters.
- An identifier may only allow characters such as lower are letters *a to z*, uppercase *A to Z*, decimal digits 0 to 9, and underscore '_'.
- Identifiers are classified into two categories. They are **basic identifier** and **extended identifier**.

3.6.2 Basic Identifiers

- The constructive rules for basic identifiers are :
  1. It can contain only alphabet letters (Lower case or Upper case), decimal, numbers and under case.
  2. It must begin with an alphabet letter and should not end with under case.
  3. It allows lower case and upper case letters.
  4. No spaces are allowed inside an identifier.
  5. No adjacent under case are allowed.
  6. It may be of any length as desired. But their size limitation on some tools, which should not exceed 16 characters.
  7. Reserved or keywords should not be used.
- These constructive rules can, at times, be overlay restrictive. For example, digital system data books often designate a part name with leading numerical, such as 74HC00. Digital system data books also often designate active-0 signal state with a leading slash, such as/ALARM or a trailing dash – such as ALARM –. These sample names are not legal VHDL names.
- Inability to express these names as VHDL impeder describing Existing designs in VHDL and often leads to cumbersome and awkward. Hence, VHDL – 93 provides an enhanced set of rules constructing identifiers that includes both the **basic identifiers** and **extended identifiers**.

3.6.3 Extended Identifier

- More general identifiers are called extended identifiers.
- Rules of constructing extended identifiers in VHDL-93 are listed as under :
  1. It can be of any length, as many characters as desired.
  2. It must be delimited by leading and trailing back-slashes.
  3. The allowed characters are any graphic character, which includes all the characters allowed for VHDL-93 basic identifiers plus special characters, such as dash "–", asterisk "*", A circumflex "A", and e umlaut "e".
  4. Within the endasing backslashes and graphic characters can apply in any order, except that a back slash used as part of an extended identifier must be denoted by two backslashes.
  5. An extended identifier is **case sensitive**, meaning that there is a difference uppercase and lowercase letters.

3.7 VHDL DATA TYPES

3.7.1 Introduction

- VHDL provides a variety of data types and operators in the package STANDARD that supports the methodology of top-down design using abstractions of hardware in early version of designs.
- Besides standard types and operations, VHDL supports user defined data types that can be included in own user package.

- The concept of type is very important, when describing data in a VHDL model. The type of a data object defines the set of venders that an object can assume as well as the set of operation that can be performed.
- Every data object in VHDL can hold a value that belongs to a set of values, specified by using a **type declaration**.
- A type is a name that has associated with it is a set of values and a set of operations. Certial types, and operations that can be performed on objects of these types, are predefined in the language.
- The language also provides the facility to define new types using type declarations and also to define a set of expressions on these types by writing functions that return values of this now type.

3.7.2 Definition

- Data type is a classification objects or items or data that defines the possible set of values, which the objects or items belonging to that type may assume.
- **Data type can be defined as a template as blue print that describes the behavior of the statement that the output of its type supported.**
- The class describes the contents of the objects that belong to it. It describes an aggregate of data fields, called instance variables, and defines the operations, called a method of object.

3.7.3 Major Categories

- The four major categories of data types in VHDL are as under :
  - (1) Scalar types
  - (2) Composite types
  - (3) Access types
  - (4) File types

3.7.4 Classification

- The classification of data types in VHDL is as shown in Fig. 3.1.

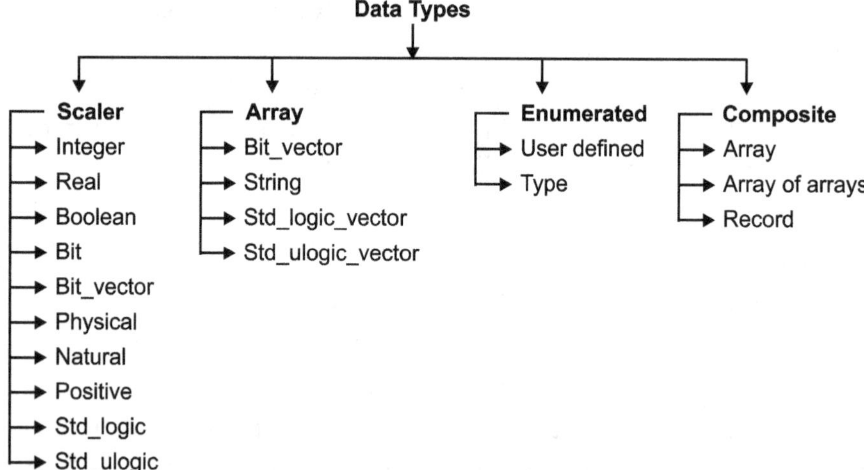

Fig. 3.1 : Data Types

3.7.5 Scalar Types

- The values belonging to the scalar type are order of sequentially, that is rotational operators can be used an these values. For example, BIT is a scalar type and the expression '0' < '1' is valid and has the value TRUE. There are four kinds of scalar types. They are as under :
  - 1. Enumeration,
  - 2. Integer,
  - 3. Physical,
  - 4. Floating point
- Integer types, floating point types, and physical types are classified as numeric types, since the values associated with these type are numeric.
- Enumerated and integer types are called *discrete types*, since these types have discrete values associated with them.

- Every value belonging to an Enumerated type, integer type, or a physical type has a position number associated with it. This number is the position of the value in the ordered list of values belonging to that type.

3.7.6 Enumerated Types

- **An enumerated type of declaration defines a type that has a set of user defined values consisting of identifiers and character literal.**
- MVL is an Enumerated type that has the set of order values ; 'U', '0', '1' and 'Z'.
- ARTH_OP is a subtype of the base type MICRO_OP and has a range constraint specified to the from ADD to DIV, that is, the values ADD, SUB, MNL, and DIV being to the subtype ARTH_OP.
- A range constraints can be specified an object declaration as shown in the signal declaration for clock. Here the value of signal clock is restricted to '0' or '1'.
- The order of values appearing in an Enumerated type declaration defines the local order for the values, that is when using reformal operators, or due is always less than a value that appears to it right in the order.
- For example, in the MICRO_OP type declaration, STORE, <DIV is True, and SUB > MNL is False.
- If the same literal is used in two different Enumerated type declarations, the literal is said to be *overloaded*. Then whenever such a literal is used, the type of the literal to determine from its surrounding contact.
- The predefined Enumerated types of the longer use CHARACTER, BIT, BOOLEAN, and SEVERITY_LEVEL.
- Values belonging to the *CHARACTER* constitute the 128 characters of the ASCII character set. These values are called *character* electrode and are always written between two single quotes ("....").
- The predefined type BIT has the literals '0' and '1' while type BOOLEAN has the literals FALSE and TRUE. Type SEVERITY_LEVEL has the values NOTE, WARNING, ERROR, and FAILURE; this type is typically used in assertion statements.

3.7.7 Integer Type

- **An integer type defines a type whose set of values fall within a specified integer range.**
- **INDEX is an integer type that includes the integer value from 0 through 15.**
- DATA_WORLD is a subtype of WORLD_LENGTH that includes the integer values ranging from 15 throughout 0. The portion of each value of an integer type is the value itself.
- In the type of declaration of WORLD_LENGTH, value 31 is at position 31, value 14 is at position at 14, and so on.
- In the declaration of MY-WORLD, the position number of values 4, 5, and 6, is 4, 5 and 6 respectively.
- Construct this with the position number of elements and enumerated type; the position number in case of integer types does not refer to the index of the element in the range, but to its numeric value itself.
- The bounds of the range for an integer type must be constant or locally static expressions; a **locally static expression** is an expression that computes to a constant value at compile time to **globally static expression** is a constant value of elaboration timed.
- Values belonging to the only predefined integer type of the language, its range is implementation dependent, but must at least cover the range; $-2^{31} - 1$ to $-2^{31} - 1$.
- Values belonging to an integer type are called *inter literals*. Examples of integer literals are 56349 6E2 098_71_28.

3.7.8 Floating Point Type

- A floating point type has a set of values in a given range of numbers.
- LENGTH is a variable object of the REAL_DATA that has been constrained to take real values in the range 0.0 through 15.9 only.
- The range constraint was specified in the variable declaration itself. Alternately, it is possible to declare type and then use. This subtype is the variable declaration.

- The range bounds specified in a floating point type declaration must be constant or locally static expression.

- Floating point literals are values of a floating point type. These can differ from integer literals by the presence of the dot (·) character. Thus 0 is an integer literal, while 0·0 is a floating point literal.

- Floating point literals can be expressed in exponential form, the exponent represents a power of ten and the exponent value must be an integer.

- Inter and floating point literals can also be written in a base other than 10 (decimal). The base can be any value between 2 and 16. Such literals are called *based literals*. The exponent represents a power of the specified base. The base and exponent values in a based literal must be in decimal notation.

- The only predefined floating point type is REAL. The range of REAL is again implementation dependent, but it must at least cover the range I · OE38 to + I · OE38 and it must be able for at least six decimal digits of precision.

3.7.9 Physical Type

- A physical type contains values that represent measurement of some physical quantity like time, length, voltage and current. Values of this type also express integer multiples of a base unit.

- CURRENT is defined to be a physical type that contain values from 0 nA to 1 nA.

- The base unit is a nA, while all others are derived units. The position number of a value is the number of base units represented by that value.

- Values of a physical are called physical literals. Physical literals are written as an integer literal followed by the unit name.

- The only predefined physical type is TIME and the range of base values, which again is implementation dependent must at least be $- (2^{31} - 1)$ to $+(2^{31} - 1)$. The declaration of type TIME appears in package STANDARD.

3.7.10 Composite Types

- A composite type represents a collection of values. There are two composite types: an array type and a record type. An array type represents a collection of values all belonging to a single type. On the other hand, a record type represents a collection of values that may belong to same or different types.

- An object belonging to a composite type represents a collection of subobjects, one for each element of the composite type. An element of a composite type could have a value belonging to either a scalar type, a composite type, or an access type. A composite type may be defined to represent an array of an array of records. This provides the capability of defining arbitrarily complex composite types.

1. **Array Type :**

- An object of an array type consists of elements that have the same type. ADDRESS_BUS is a one-dimensional array object that consists of 64 elements of type BIT.

- ROM_ADDR is a one-dimensional array object that consists of 126 elements, each element being another array object consisting of 8 elements of type MVL. Hence an array of arrays is created. Elements of an array can be accessed by specifying the index values into the array.

- ADDRESS_BUS(26) refers to the 27th element of ADDRESS_BUS array object. ROM_ADDR(10)(5) refers to the value (of type MVL) at index 5 of the ROM_ADDR(10) data object (of type DATA_WORD); DECODER(5,2) refers to the value of the element at the 2nd column and 5th row of the two-dimensional object.

- The language allows for an arbitrary number of dimensions to be associated with an array. It also allows an array object to be assigned to another array object of the same type using a single assignment statement.

- Assignment can be made to an entire array, or to an element of an array, or to a slice of an array.
- The language also allows array types to be unconstrained, that is, the number of elements in the array is not specified in the type declaration. Instead, the object declaration for that type declares the number of elements of the array.
- A subtype declaration may also specify the index constraint for an unconstrained array type. A subprogram parameter may be an object specified to be of an unconstrained type, the constraint is specified by the actual parameter passed in during the subprogram call.
- STACK_TYPE is defined to be an unconstrained array type that specifies the index of the array as an integer type STACK_TYPE, also specifies an index constraint.
- The constant ALU_TIMING specifies the timing for a two-operator ALU, where the operators could be ADD, SUB, or MUL. An ALU that performs ADD and SUB operations has a delay of 20 ns.
- The declaration for ALU_TIMING is a special case of a constant declaration where no constraint need be specified for the unconstrained array type, since the size of the constant object is determined from the number of values in the constant.
- There are two predefined one-dimensional unconstrained array types in the language, STRING and BIT_VECTOR. STRING is an array of characters while BIT_VECTOR is an array of bits.
- A value representing a one-dimensional array of characters is called a ***string literal***. String literals are written by enclosing the sequence of characters within double quotes ("...").
- A string literal can be assigned to different types of objects, to a STRING type object or to a BIT_VECTOR type object. The type of a string literal is determined from the context in which it appears.
- A string literal that represents a sequence of bits (values of type BIT) can also be represented as a ***bit string literal***. This sequence of bits, called ***bit strings***, can be represented either as a binary value, or as an octal value, or as a hexadecimal value. The underscore character can again be freely used in bit string literals for clarity.
- There are many different ways to assign values to an array object. An ***aggregate*** is a set of comma separated elements enclosed within parenthesis.

2. Record Type :

- An object of a record type is composed of elements of same or different types. It is analogous to the record data type in PASCAL and the struct declaration in C.
- Values can be assigned to a record type object using aggregates. Thus values can be assigned to a record type object from another record type object of the same type using a single assignment statement.
- Each element of a record type object can also be assigned individually by using selected names.
- Aggregate values can be assigned to record type objects using both positional and named associations, since an aggregate does not specify its type, it can be either an array or a record aggregate, depending on its usage.
- If CAB is an array type object, the aggregate is treated as an array aggregate; on the other hand, if CAB is a record type object, the aggregate is a record aggregate where others refers to all the elements in the record.
- If type-name is an unconstrained array type, a different READ procedure is implicitly declared,
- A file cannot be opened or closed explicitly and values within a file can only be accessed sequentially.
- A test bench that reads vectors from a file,. "fadd.vec", applies these vectors to the test component, a one-bit full adder, and then writes back the results into another file, "fadd.out".
- VEC_FILE is an input file and contains 3-bit strings, and RESULT_FILE is an output file into which 2-bit strings are written. Input vectors are read one at a time until end of file is reached. Each vector is applied and then the process waits for the full-adder circuit to stabilize before sampling the adder output value which is written to the output file.

- One file type, TEXT, is predefined in the language; this file type represents a file consisting of variable length ASCII strings.
- The assertion statement when executed simply stops the simulation run since the assertion condition is always false. This statement is not necessary if the VHDL simulator provides a facility to control the simulation run.
- An access type, LINE, is also provided to point to such strings. Operations to read and write data from a single line are provided. Operations to read and write entire lines are also provided. The definitions for all these types and operations appear in the predefined package, TEXTIO.

3. **Access Types :**
- Values belonging to an access type are pointers to a dynamically allocated object of some other type. They are similar to pointers in Pascal and C languages. Access type provide access to objects of a given type via pointers.
- PTR is an access type, whose values are addresses that point to objects of type MODULE. Every access type may also have the value null, which means that it does not point to any object.
- Objects of an access type can only belong to the variable class. When an object of an access type is declared, the default value of this object is null.
- Objects that access types point to, can be created using **allocators**, which provide a mechanism to dynamically create objects of a specific type.
- The new in this assignment causes an object of type MODULE to be created and the pointer to this object is returned.
- The values of the elements of the MODULE record are the default values of each element; these are 20 (the leftmost value of the implied subtype) for the SIZE element.
- TIME'LEFT for the CRITICAL_DLY element, and the value 0 (this is PIN_TYPE'LEFT) for the NO_INPUTS and NO_OUTPUTS elements. Initial values can also be assigned to a newly created object by explicitly specifying the values.
- Access types are useful in modeling high-level behavior, especially that of regular structures like RAMs and FIFOs, where pointers can be used to access objects one at a time in a sequence.

4. **File Types :**
- File types provide access to objects that contain a sequence of values of a given type. Objects of file types represent files in the host environment, which provide a mechanism by which a VHDL design communicates with the host environment.
- The **type-name** is the type of values contained in the file. A file of type VECTORS has a sequence of values of type BIT_VECTOR; a file of type NAMES has a sequence of strings as values in it. A file is declared using a file declaration.
- The **string-expression** is interpreted by the host environment as the physical name of the file.
- The mode of a file, in or out, specifies whether it is an input or an output file, respectively. Input files can only be read while output files can only be written to.
- VEC_FILE is declared to be a file that contains a sequence of bit vectors and it is an input file. It is associated with the file "/usr/home/jb/uart/div.vec" in the host environment.
- A file belongs to the **variable** class of objects. However, a file cannot be assigned values using a variable assignment statement. It can be read, written to, or tested for an end-of-file condition, only by using special procedures and functions that are implicitly declared for every file types.
- The file is declared by a file declaration. There are two other objects that are explicitly declared. These are indices of a for loop and it generates statement.
- The file type is used to access file on disk. It is used only in test bench, in fact, file type cannot be implemented in hardware.
- In orders to use file type, we shall include the packages that contains all procedure and functions that allow you to read from and write to the form called text files.

- Input ASCII files are handled as file of lines where a line is a string terminated by a carriage return. Package declares a type used to hold a line read from an input file and a line to write to an output file.

3.8 PREDEFINED DATA TYPES

3.8.1 Introduction

- VHDL consists of two packages, named as STANDARD and TEXTIO. The STANDARD package of data type is induced in all VHDL source files.
- TEXTIO package defines types and operations for communications with a standard programming environment (terminal and file input/output).
- VHDL has a set of standard data types that is, **predefined/built**. It is also possible to have user defined data types and subtypes.
- The TEXTIO package is not needed for synthesis, and therefore VHDL compiler does not support.
- The STANDARD package is a subset of the IEEE VHDL package.

3.8.2 Classes

- Some of the predefined data types in VHDL are as under :

  1. Bit
  2. Bit_Vector
  3. Boolean
  4. STD_LOGIC
  5. STD_LOGIC_VECTOR

- The STD_LOGIC and STD_LOGIC_VECTOR data types are not built in VHDL data types, but are defined in the standard logic 1164 package of IEEE library. We, therefore, need to include this library in VHDL code and specify that the STD_LOGIC_1164 package must be used in a scalar to use the STD_LOGIC data type.

3.8.3 BIT Data Type

- A BIT data types predefined in the standard package as in enumerated data type with only two binary values '0' and '1'.
- When assigning a value of 0 or 1 to a BIT in VHDL code, the 0 or 1 must be enclosed in single quotes, '0' or '1'.
- Logical operations, such as AND can take and return BIT values.
- Convert a BOOLEAN value to a BIT value as follows :

  if (BOOLEAN_VAR) then

      BIT_VAR : = '1';

  else

      BIT_VAR : = '0';

  end if;

- BIT is used for single line signal.

3.8.4 BIT_VECTOR Data Type

- A BIT_VECTOR data type is predefined in the standard package as one dimensional array type with each element being of the BIT data type. It is used for multi-line signal, which may be 4.
- The BIT_VECTOR data type represents an array of BIT values.
- BIT-VECTOR data type is the vector version of the BIT type consisting of two or more bits. Each bit in a BIT-VECTOR can only have value 0 or 1.
- When assigning a values in a BIT-VECTOR, the value must be enclosed in double miles, e.g. "1011" and the number of bits in the value must match the size of the BIT_VECTOR.

3.8.5 BOOLEAN Data Type

- A BOOLEAN data type is predefined in the standard packages as an enumerated data types with two possible values as False and True.

- The BOOLEAN data type is actually an Enumerated type with two values, FALSE and TRUE, where FALSE < TRUE.
- Logical functions, such as equality (=) and comparison (<) function can take and return a BOOLEAN values.
- Convert a BIT value to a BOOLEAN value as follows:
  BOOLEAN_VAR := (BIT_VAR = '1');

3.8.6 STD_LOGIC Data Type

- The STD_LOGIC data type can have the values x, 0, 1, or z. There are other values that this data type can have, but the other values are not synthesizable, i.e. they cannot be used in VHDL code that will be implemented on a CPLD or FPGA.
- These values have the following meanings:
  1. X – Unknown,　　　　　　　　　2. 0 – Logic 0,
  3. 1 – Logic 1,　　　　　　　　　　4. Z – High impedance (open circuit) / tristate buffer
- When assigning a value to a STD_LOGIC data type, the value must be enclosed in single quotes: 'X', '0', '1' or 'Z'.

3.8.7 STD_LOGIC_VECTOR Data Type

- It is the vector version of the STD_LOGIC data type. Each bit is the set of bits that make up the vector can have the values X, 0, 1 or Z.
- When assigning a value to a STD_LOGIC_VECTOR type, the value must be enclosed in double quotes, e.g. "1010", "ZZZZ" or "ZZ001". The number of bits in the value must match the size of the STD_LOGIC_VECTOR data type.

3.9 VHDL DATA OBJECTS

3.9.1 Introduction

- A data objects in VHDL holds a value of a specified type, created by using object declaration.
- Each value can be accessed by using its identifier or a more complex expression that refers to the object. In addition, each object has a unique data type.
- An object declaration is used to declare an object, its type and its class, and optimally assign it a values.
- Every data object belongs to either constant, variable or signal.

3.9.2 Definition

- **A data object in VHDL is a region of the storage that contains a value or a group of values.**

3.9.3 Classes

- The VHDL data object can be categorized into three following classes :
  1. Constant,　　　　　　　　　2. Variable　　　　　　　　　3. Signal.

3.9.4 Constants

- Constants are the objects, whose values cannot be changed after it is initially specified.
- An object of constant class can hold a single value of a given type.
- For example,
  constant RISE_TIME: TIME := 10ns;
  constant BUS_WIDTH: INTEGER := 8:
- The first constant declaration declares the objects RISE_TIME that can hold a value or type TIME, with the value assigned at the start of simulation in 10 ns.
- The second declaration declares a constant BUS-WIDTH type INTEGER with a value of 8.
- The value of the constant has not bean specified called as a deferred constant and it can appear only inside a package declaration. The complete constant declaration with the associated value must appears in the corresponding package body.
  constant No_of_INPUTS : INTEGER;
- By use of a constant VHDL model becomes more readable and easy to update.

- The syntax for constant is

  constant constant_name : type_name := value;
- A single value of a given type is assigned to the constant before simulation starts and this value cannot changed during the source of simulation.
- Constant names are assigned to specific values value to a type to the VHDL model better decremented and easy to update.
- Constants can be declared in all the declarations VHDL statements such as sequential and concurrent section of packages, entities, architectures, processes, subprograms and blocks.

3.9.5 Variable

- **A variable is an object with single current value which may be changed.**
- An object of variable class can hold a single value of a given type. However, in this case, different values can be assigned to the object at different times using a variable assignment statement.
- An object with only a current value, variable values or desired variables go away after synthesis.
- The variable locally stores temporally data and it is used only inside a sequential statement that means processes, function and procedures.
- The variables are used inside a process to perform action sequentially. The most common variables in VHDL are used in construction of gate, component and other pieces of digital hardware.
- The variable is visible only inside processes and subprograms.
- The variables are also used for local storage in process statements and subprograms.
- For example,

  Variable COUNT : INTEGER;
- This results in the creation of data object called COUNT, which can hold into gate values and the object COUNT is also declared to be **variable** class.

  Variable CTRL_STATUS : BIT_VECTOR (10 downto 0);

  Variable_SUM : INTEGER range 0 to 100 : = 10;

  Variable FOUND, DONE : BOOLEAN;
- The first declaration specifies a variable object CTRL_STATUS as an array of is elements, with each array element of type BIT.
- In the second declaration, an explicit initial value has been assigned to the variable sum, with a an initial value. This default value is TLEFT, where T is the object type and LEFT is a predefined attribute of a type that gives the left most value in the set of values belonging to type T.
- In the third declaration, the initial values assigned to FOUN and DONE at the start of simulation is FALSE. The initial value for all the array elements of CTRL_STATUS is '0'.
- The variable declaration syntax is as under,

  variable variable_name : type_name : initial value

3.9.6 Signal

- **Signal is an object with current values and projected value.**
- Signal is a communication medium between entities. Signal is nothing but wires that connect two or more components laying inside an IC. Thus signal represents interconnection between the parts of VHDL objects.
- An object belonging to the signal class has a past history of values, a current value, and a set of future values. Future values can be assigned statements.
- The signal object can be regarded as wires in a circuit, while variable and constant objects are analogous to their counterparts in a high-level programming language like C or PASCAL.
- Signals can be declared in entity declaration, architecture declaration and package declaration sections.

- Signal declared in package can be shared among entities and are called global signals.
- Signal holds a list of values, which include the current value of the signal and a set of future values that are to appear on the signal.
- Signal objects are typically used to models wires and flip flops, where variables and constant objects are typically used to model the behavior of the circuit. They are also used to connect entities together to form models.
- Signals declared in entity declaration section are global to only architecture c on be referenced in that architecture only.
- The projected values can be changed as many times as possible. This is most common type in design. We use signals to construct internal bus, shift registers, RAM, etc. We use digital circuits to create signals.
- For example,

    signal CLOCK : BIT;

    signal DATA_BUS : BIT_VECTOR (0 to 7);

    signal GATE_DELAY : TIME : = 10 ns;

- The first signal declaration declares the signal object CLOCK of type BIT and gives it an initial value of '0'.
- The second signal declaration declares the signal object DATA_BUS of type BIT_VECTOR with one-dimensional array of 8 bits with an initial value of '0000000'.
- The third signal declaration declares a signal object GATE_DELAY of type TIME that has an initial value of 10 ns.

3.10 VHDL OPERATORS

3.10.1 Introduction

- VHDL has a wide set of different operators, which can be divided into groups of the same precedence level (priority). Thus VHDL provides predefined operators, which are used as hardware modeling units.
- *Operators are means for constructing expressions.* The expressions are evaluated form left to right, operations with higher precedence are evaluated first. If the order should be different from the one resulting from this rule, parentheses can be used.
- The operands, connected with each other by an operator, are evaluated before the operation described by that operator is carried out.
- For some operators the right operand is evaluated only when the left operand has a certain value assigned to it.
- The operators for the predefined types are defined in the STANDARD package in the STD library. These operators are functions, which always return the same value when they are called with the same values of the actual parameters. These functions are called the *pure* function.

3.10.2 Classification

- Every language has operators, whose functions are to operate on operands and produce some result. A subset of VHDL operators can be classified into following groups :

    1. Logical operators   2. Relational operators
    3. Shift operators   4. Arithmetic operators
    5. Assignment operators   6. Miscellaneous operators

- These VHDL operators have increasing priority in serial order as mentioned above. The logical operations have lowest priority and miscellaneous operations have highest priority.
- The operations in the same category have the same procedure. The expressions are evaluated from left to right. The operations with higher priority are evaluated first parenthesis can be used to oversize the left to right evaluation.

3.10.3 Logical Operators

- The logical operators work on predefined types; Bit and Boolean type as well as one dimensional array of Bit and Boolean.
- The concept of logical operations is simple. They allow a program to make a decision based on multiple conditions. Each operand is considered a condition that can be evaluated to a true or false value. Then the value of the conditions is used to determine the overall value of the OP1 operator, OP2 operator or OP1 grouping.
- Logical operators are typically used with Boolean values.
- The logical operators defined in VHDL are as under :
  1. and   2. or   3. nand   4. nor   5. xor   6. xnor   7. not
- All these seven operators have the lowest priority, except for the operator not, which has the highest priority as miscellaneous operators.
- The results of the logical operators for the predefined types are presented in the following tables. The BIT type is represented by the value '0' and '1', while the BOOLEAN type by True and False.

Table 3.2 : Operator 'not'

A		not A	
True	'1'	False	'0'
False	'0'	True	'1'

Table 3.3 : Operator 'and'

A		B		A and B	
True	'1'	True	'1'	True	'1'
True	'1'	False	'0'	False	'0'
False	'0'	True	'1'	False	'0'
False	'0'	False	'0'	False	'0'

Table 3.4 : Operator 'or'

A		B		A or B	
True	'1'	True	'1'	True	'1'
True	'1'	False	'0'	True	'1'
False	'0'	True	'1'	True	'1'
False	'0'	False	'0'	False	'0'

Table 3.5 : Operator 'xor'

A		B		A xor B	
True	'1'	True	'1'	False	'0'
True	'1'	False	'0'	True	'1'
False	'0'	True	'1'	True	'1'
False	'0'	False	'0'	False	'0'

Table 3.6 : Operator 'nand'

A		B		A nand B	
True	'1'	True	'1'	False	'0'
True	'1'	False	'0'	True	'1'
False	'0'	True	'1'	True	'1'
False	'0'	False	'0'	True	'1'

Table 3.7 : Operator 'nor'

A		B		A nor B	
True	'1'	True	'1'	False	'0'
True	'1'	False	'0'	False	''0'
False	'0'	True	'1'	False	'0'
False	'0'	False	'0'	True	'1'

Table 3.8 : Operator 'xnor'

A		B		A xnor B	
True	'1'	True	'1'	True	'1'
True	'1'	False	'0'	False	''0'
False	'0'	True	'1'	False	'0'
False	'0'	False	'0'	True	'1'

Table 3.9 : Relational operations

=	Equality
/=	Inequality
<	Ordering „less than"
<=	Ordering „less than or equal"
>	Ordering „greater than"
>=	Ordering „greater than or equal"

- The operators, nand and nor are not associative. Therefore, the syntax at nand or nor operators is illegle. For example,

        c : a and b; --variables

        z < = a and b and c; -- signals

3.10.4 Relational Operators

- The relational operators allow checking relation between operands, i.e. to state whether they are equal, not equal or are ordered in a way defined by operator as tabulated in Table 3.9. Both operands must be of the same type, and the result received is always of the Boolean type.

- The operators, equality and inequality are predefined for all types available in the language except the file type. For other relations the operands must be of a scalar type or one-dimensional array types.

- The equality operator returns the value *TRUE* only when both operands have the same values, and *FALSE* when the values are different.

- The inequality operator returns the value *TRUE*, when the operators are different and FALSE when they are equal. There are certain rules that are used to compare operands depending on their type: in case of the scalar type, the operand values are equal only when the values are the same. Two values of the composite type are equal only when each value of the left operand corresponds to the value of the right operand and vice versa.

- In the record, the corresponding elements have identical identifiers, and in the array the corresponding elements are those which appear at the same positions of arrays.

- In particular two null arrays of the same type are always equal. The operators: <, <=, >, and >= return the TRUE logical value only when the condition in the given relation is met, otherwise the FALSE value is returned.

- The relational operators are used to check the conditions.

- The equality '=' and inequality '/=' operations are to predefined for all data types, except file type. The remaining four operations, such as '<', '< =', '>', '> =', are predefined to the right. These operators return the TRUE logic value, where the condition in the given relation is met.

3.10.5 Shift Operators

- The shift operators are defined for the one-dimensional array with the elements of the type BIT or BOOLEAN.

- For the shift operator an *array* is the left operand L and *integer* is the right operand R. The right operand represents the number of positions the left operand should be shifted.

- As the result of shifting, the value of the same type as the left operand is returned. Table 3.10 below shows predefined shift operators.

- The *sll* returns the value of the L left operand, after it has been shifted R number of times. If R is equal to 0 or L is the null array, the left operand L is returned.

- The single left logical operation replaces L with concatenation of the rightmost L'Length 1 elements of L and a single value T'Left, where T is the element type of L.

Table 3.10 : Shift operators

sll	Shift left logical
srl	Shift right logical
sla	Shift left arithmetic
sra	Shift right arithmetic
rol	Rotate left logical
ror	Rotate right logical

- If R > the single shift operation is repeated R number of times. If R is a negative number, the value of the expression L *srl* -R is returned.

- The operator *srl* returns the value of the left operand L after it has been shifted to right R times. In case when R is equal to 0 or L is the null array, the left operand L is returned.

- The single shifting operation replaces the operand L with the concatenation of the leftmost L'Length -1 elements of L and a single value T'Left, where T is the element type of L.
- If R is a negative number then the value of expression L sll-R is returned, otherwise the single shift operation is repeated R number of times.
- The operator sla returns the value of the left operand L after it has been shifted to the left R number of times. In case when R is equal to 0 or L is the *null array* the left operand L is returned.
- The single shift operation replaces L with concatenation of the rightmost L'Length is a negative number, the value of the expression L sla -R is returned, otherwise the single shift operation is repeated R number of times.
- The operator sra returns the value of the left operand L after it has been shifted to the right R number of times.
- In case when R is equal to 0 or L is the *null array*, the left operand L is returned. The single shift operation replaces L with a value, which is the result of concatenation whose left argument is the leftmost L'Length -1 elements of L and whose right argument is L'Left. If R is a negative number, the value of the expression L sla -R is returned, otherwise the single shift operation is repeated R number of times.
- The operator rol returns the value of the L left operand after it has been rotated to the left R times. In case when R is equal to 0 or L is the *null array* the left operand L is returned.
- Each of the operands takes an array of BIT or BOOLEAN as the left operand and an integer value as the right operand and performs the specified operation.
- If the integer value is a negative number, then the opposite action is performed, i.e. a left shift as rotate becomes a right shift or rotate respectively and vice-versa.

3.10.6 Arithmetic Operations

1. Introduction :
- The arithmetic expressions must be evaluated according to the internal sequence of operations. There are several methods to specific this operation.
- The basic arithmetic operations are addition, subtraction, multiplication and division.

2. Classification :
- Apart from the four basic arithmetic operations, there are number of arithmetic operations.
- The arithmetic operations are classified into ten categories as under :

1. Exponential,	2. Remainder,
3. Modulus,	4. Division,
5. Multiplication,	6. Subtraction,
7. Addition,	8. Absolute value,
9. Unary minus	10. Unary plus.

- Subtraction is nothing but the addition of positive and negative numbers. Thus subtraction can be performed by addition operation. Similarly division is nothing but the multiplication of one number and reciprocal of other number. Thus division can be performed by multiplication operation. Therefore, in VHDL only adding and multiplying operators are defined as arithmetic operators.

3. Adding operators :
- Adding operators consist of addition, subtraction and concatenation. The adding operators are shown in Table 3.11 below.
- The addition (+) and subtraction (–) operators perform mathematical operations, and their operands can be of any numeric type.
- The concatenation (&) operator is defined for elements of one-dimensional arrays. In concatenation, the following situations can take place.

Table 3.11 : Adding operators

+	Addition
-	Subtraction
&	Concatenation

- When both operands are one-dimensional arrays of the same type, the concatenation connects the two arrays into one.

- The new array contains the elements from both arrays. The direction of the new array is the same as the direction in the array of the left operand.

- In case, when both operands are **null arrays**, the direction of the right operand is assumed. When one of the operands is a one-dimensional array and the second operand is a scalar of the same type as the elements of that array, then the result of the concatenation is the same as in the first point. However, in this case the second operand is treated as a one-dimensional array which contains only one element.

- In case when both operands are of the same scalar type, then the result of concatenation is one-dimensional array with elements of the same types as the operands.

4. Multiplying operators :

- The multiplication and division operators are predefined for all integers, floating point numbers. Under certain conditions, they may be used for operations on physical type objects as well.

- The mod and rem operators, on the other hand, are defined only for the integers. When mod and rem operators are used, then both the operands and the result are of the same integer type. The multiplying operators are shown in Table 3.12.

Table 3.12 : Multiplying operators

*	Multiplication
/	Division
mod	Modulus
rem	Remainder

5. Sign operators :

- Sign operators are unary operators, i.e. have only one, right operand, which must be of a numeric type. The result of the expression evaluation is of the same type as the operand. There are two sign operators such as identify (+) and negation (–).

- When (+) sign operator is used, the operand is returned unchanged, but in case of ( ) sign operator the value of operand with the negative sign is returned.

Table 3.13 : Sign operators

+	Identity
–	Negation

- Because of the lower priority, the sign operator in the expression cannot be directly preceded by the multiplication operator, the exponentiation operator (**) or the **abs** and **not** operators.

- When these operators are used then sign operator and its operand should be enclosed in parentheses.

3.10.7 Miscellaneous Operators

- The two miscellaneous operators defined in VHDL are exponentiation (**) and absolute value (abs) as shown in Table 3.14.

Table 3.14 : Miscellaneous operators

**	Exponentiation
abs	Absolute value

- The *exponentiation* operator has two operands. This operator is defined for any integer or floating point number. The right operand (exponent) must be of integer type.

- When the exponent is the positive integer, then the left operand is repeatedly multiplied by itself.

- When the exponent is the negative number, then the result is a reverse of exponentiation with the exponent equal to the absolute value of the right operand (Example 9). If the exponent is equal to 0 the result will be 1.

- The *abs* operator has only one operand. It allows defining the operand's absolute value. The result is of the same type as the operand.

- The operator not is classified as a miscellaneous operator only for the purpose of defining procedure. Otherwise, it is defined as a *logical operator*.

3.10.8 Unary Operators

- Unary operation is an operation with only one operand, i.e. a single input. This is in contrast to the binary operations, which use two operands. An example is the function f: A – A, where A is a set. The function f is a binary operation on A.
- Common notations are predefined notation (e.g. +, –), postfix notation (e.g. actuarial n'1), functional notation (e.g. sin y or sin (x)), and subscripts (e.g. transport A^T). Other notations also exist as well.

3.11 VHDL DELAY TYPES

3.11.1 Introduction

- In VHDL, the designer has the possibility to perform a signal assignment after certain amount of time, implementing the delays in the assignment.
- The delay mechanism allows introducing propagation times of described systems.
- Delays are specified in signal assignment statement. It is not allowed to specify delays in variable statements.
- Delay mechanism can be applied to signals only, which are applied to systems. Delays are not synthesizable.

3.11.2 Definition

- *Delay is mechanism allowing introducing timing parameters of specified systems.*

3.11.3 Types of Delays

- There are basically two delay mechanisms available in VHDL. They are as under :
    1. Inertial delay (default),
    2. Transport delay.

3.11.4 Inertial Delay

- If the delay mechanism in VHDL is not specified, then by default it is inertial delay.
- The inertial delay is defined using the reserved word **inertial** and is used to model the devices, which are inherently inertial. In practice this means, that impulses shorter than specified switching time are not transmitted.
- The inertial delay specification may contain a **reject** clause. This clause can be used to specify the minimum impulse width that will be propagated, regardless of the switching time specified.
- As inertial delay is the default delay, the reserved word inertial can be omitted and you may simply call it as delay.
- The inertial delay models the delay introduced by an analog port, which means, it is analogous to the delay in devices that respond only if the signal value persists on their inputs for a given amount of time.
- The inertial delay model is the default delay implemented in VHDL because it's behavior is very similar to the delay of the device. The delay assignment syntax is as under :

> b <= a **after** 20 ns;

- In this example *b* take the value of *a after* 20 ns second of inertial delay. This means that if a value varies faster than 20 ns *b* remain *unchanged*. This simulation should clarify the concept.
- At simulation start *a* and *b* are 0; a change from 0 to 1, and *b* change its value after *20 ns*; then *a* changes value going to *0* and then to *1* in *10 ns*. This delay is *less than* inertial delay of 20 ns. So *b* remains *unchanged*.
- Thus inertial delay is often found in switching circuits. It is the default delay in digital circuits. This delay is often used to filter unwanted spikes and transients on the signals.

- The inertial delay represents the time for which an input must be stable before the value is allowed to propagate too the output. In addition, the value appears at the output after the specified dealy.

3.11.5 Transport Delay

- The transport delay is defined using the reserved word **'transport'** and is characteristic for transmission lines.

- Transport delay models the delays in hardware that do not exhibit any inertial delay. This delay represents pure propagation delay, i.e, any changes on an input is from ported to the output, no matter how small, after the specified delay.

- To use a transported delay model, the keyword transport must be used in a signal assignment statement.

- New signal value is assigned with specified delay independently from the width of the impulse in waveform, i.e. the signal is propagated through the line.

- The transport delay is not the default delay implemented in VHDL and *must be specified*. This delay model is useful to describe delay line, PCB delay, wire delay. The delay assignment syntax is:

      b <= **transport** a **after** 20 ns;

- In this example *b* take the value of *a* after 20 ns second of *transport delay*. This means that no matter how fast *a* changes his value, *b* will follow the behavior of *a* after the amount of time specified in the delay statement. This simulation should clarify the concept.

- The simulation is the same as the inertial delay. In this case the value of *b* follow the value of *a* after 20 ns even when a has a glitch of 10 ns.

- VHDL transport and inertial delay model allow the designer to model different type of behavior on VHDL hardware implementation.

- The inertial and transport delays are very useful in test bench modeling, such as RAM, ROM, and peripheral.

3.11.6 Delta Delay

- Delta delay is a very small delay, which does not correspond to any real delay and actual simulation time does not advance.

- This delay models hardware where a minimal amount of time in needed for a change to occur, for example, in performing zero delay simulation.

- Delta delay allows for ordering of events that occur at the same simulation time during a simulation.

- Each unit of simulation time can be considered to be composed of an infinite number of delta delays. Therefore, an event always occurs at a real simulation time plus an integral multiple of delta delays.

- In VHDL simulation, all signal assignments occur with some infinitesimal delay, known as delta delay. Technically, delta delay is of no measurable unit, but from a hardware design perspective one should think of delta delay as being the smallest time unit one could measure, such as femtosecond.

- A delta or delta cycle is essentially an infintesimal, but quantized, unit of time. The delta delay mechanism is used to provide a minimum delay in a signal assignment statement, so that the situation cycle described earlier can operate correctly when signal assignment statements do not include extalicity specified delays.

- All active processes can execute in the same simulation cycle. Each active process will suspend at wait statement.

- When all processes are suspended, simulation is advanced the minimum time necessary to that some signals each take on their new values.

- Processes the determine, if the new signal values satisfy the conditions to process from the wait statement at which they are suspended.

3.12 VHDL DESIGN UNITS/ELEMENTS

3.12.1 Introduction

- A design unit may be the entire file or there may be more than one design unit in a file. No less than a design unit may be in a file.
- Any design unit may contain a context clause as its initial part. The context clause of a primary unit applies to all of the primary units corresponding secondary units. Architectures and package bodies are the secondary units. Subprograms are not library units and must be inside entities, architectures or packages.
- The analysis, compilation, of a design unit results in a library unit in some design library. Predefined libraries typically include but are not limited to: STD, IEEE and WORK. WORK is the default user library.

3.12.2 Definition

- **The primary unit, which is used to design VHDL models, is called a design unit**. Sometimes unit is also referred to as element or component.
- A design unit is an entire file and it contains context clause as an initial part.

3.12.3 Classification

- The VHDL models are generally built from different kinds of design units. They are classified as under :

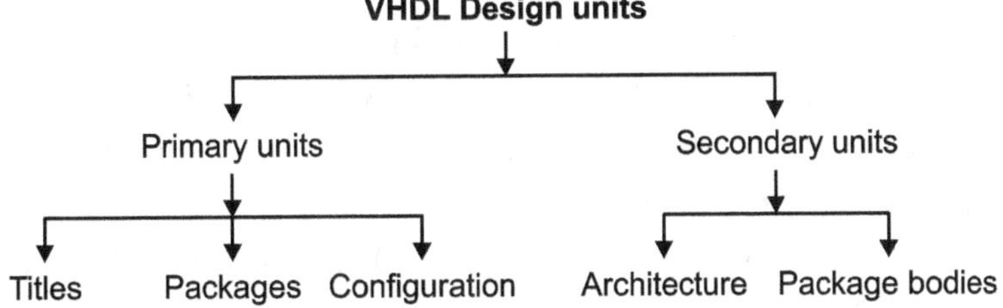

- Thus, there are five different classes of design units in VHDL as mentioned below :
    1. Entity declarations,
    2. Architectures declarations,
    3. Package,
    4. Package bodies,
    5. Configurations.
- Entities and architectures are the only two architecture basic units that have in any VHDL design description.
- Package and configuration are optional.

3.13 ENTITY DECLARATION

3.13.1 Definition

- An entity is the most basic building block in VHDL design unit. So, entity acts as a black box which gives the external view of the VHDL design.
- **An entity is the declaration of the interface between a VHDL design and its external environment.**
- An entity may also specify the declarations and statements that are part of the design entity. A given entity declaration may be shared by many design entities, each of which has a different architecture. Thus, an entity declaration can potentially represent a class of design entities, each having the same interface.

3.13.2 Simplified Syntax

```
entity entity_name is
    generic (generic_list);
    port (port_list);]
end entity entity_name;
```

3.13.3 Description

- A design is described in VHDL using the concept of design entity. A design entity is split into two parts, each of which is called a *design unit* in VHDL.

- The *entity declaration* represents the external interface to the design entity. The *architecture* body represents the internal description of the design entity; its behaviour, its structure, or a mixture of both.

- An entity specifies the interface between the specified design, formally called a design entity and the environment in which it operates. The entity describes the designates interface to the external circuit. It is equivalent to b in configuration of an IC. We can think of on entity declaration as corresponding to chip package.

- The entity declaration defines the external view of the entity; i.e. description of the input and output ports of the design. Each port in the port list must be given a name, data flow direction and a type.

- The name of the design is just an arbitrary label invented by the user. It does not correspond to a name predefined in a VHDL component library.

- The entity declaration includes the name of the entity and a set of *port* declarations. A port may correspond to a pin on an IC, an edge connector on a board, or any logical channel of communication with a block of hardware.

- Each port declaration includes the name of one or more ports, the direction that information is allowed to flow through the ports *in*, *out* or *inout*, and the data type of the ports e.g., STD_LOGIC.

- The data type of a port defines the set of values that may flow through the port. The ports are of type STD_LOGIC, which is found in package STD_LOGIC_1164 in library IEEE.

- The concept of data type is borrowed by VHDL from the world of software. It allows the VHDL compiler to ensure that the design is at least reasonably robust before beginning simulation.

- The entity declaration is terminated by the VHDL keyword end. We indulge in a little programming robustness by adding the name of the design entity after the end keyword.

- Including the name of the design entity is particularly relevant in large descriptions, where the port list may extend over many screens or pages to be reminded of the name of the design entity whose end we are looking at, lest we forget.

- The name of the architecture body is just an arbitrary label invented by the user. It is possible to define several alternative architecture bodies for a single design entity, and the only purpose of the architecture name is to distinguish between these alternatives. architecture, ofand is are VHDL keywords. Note that when we define an architecture, we have to tell the VHDL analyzer that the architecture corresponds to the design entity.

- The VHDL keyword *begin* denotes the end of the architecture declarative region and the start of the architecture statement part. In this architecture, there is but one statement, and all the names referenced in this statement are in fact the ports of the design. Because all of the names used in the architecture statement part are declared in the entity declaration, the architecture declarative part is empty.

- The architecture contains a concurrent signal assignment, which describes the function of the design entity. The concurrent assignment executes whenever one of the four ports change value.

- The architecture is terminated by the VHDL keyword *end*. Once again, we reference the architecture name at the end of the architecture body for the same reason as we did with the entity.

- Usually, architecture bodies require sigificantly more code than entity declarations, hence repeating the name of the architecture is even more relevant.

- Similar to many programming languages VHDL supports comments, which are not a part of VHDL design, but allow the user to make notes referring to the VHDL code, usually as an aid to understanding it.

- Two hyphens mark the start of a comment, which is ignored by the VHDL compiler. A comment can be on a separate line or at the end of a line of VHDL code, but in any case stops at the end of the line.
- The entity statement declares the design name, an identifier item in the syntax example. In addition, it also defines generic parameters and ports of the designer entity.
- The generic parameters provide static information like timing parameters as bus width to a design. Ports provide communication channels between the design and its environment. For each port, its mode, i.e. data flow and types, are defined.

3.14 ARCHITECTURE DECLARATION

3.14.1 Introduction

- The entities are interconnected and communicated through signal. Each entity is associated with one or more architectures. There can be no architecture without an entity.
- The architecture in VHDL specifies behavior, function, interconnections, relationship between inputs and outputs of an entity.
- The architecture build only from processes end, simple signal assignments is called *behivioural* architecture. The architecture composed from other entities referred to as components is called *structural architecture*.

3.14.2 Definition

- **A body associated with an entity declaration to describe the internal organization or operations of a design entity is called an architecture.**
- The architecture is a module used to define how entity behaves or what it is composed of.

3.14.3 Description

- The architecture body contains the internal description of the entity, e.g., as a set of interconnected components that represents the structures of the entity or as a set of component or sequential statements that represent the behavior of entity.
- The architecture assigned to an entity describes internal relationship between input and output ports of the entity. It consists of two parts, declarations and concurrent statements.
- The declaration part of an architecture may contain declarations of *types*, *signals*, *constants*, subprograms (*functions* and *procedures*), *components*, and *groups*.
- Concurrent statements in the architecture body define the relationship between inputs and outputs.
- This relationship can be specified using different types of statements namely *concurrent signal assignment, process statement, component instantiation, concurrent procedure call, generate statement, concurrent assertion statement and block statement.* It can be written in different styles: structural, dataflow, behavioral (functional) or mixed.

3.14.4 Simplified Syntax

- The simplified architecture in VHDL is as under :

architecture architecture_name **of** entity_name **is**

architecture_declarations

    begin

        concurrent_statements

    end [**architecture**] [architecture_name];

- The *architecture* body describes only the expected functionality *(behavior)* of the circuit, without any direct indication as to the hardware implementation. Such description consists only of one or more *processes*, each of which contains sequential statements

3.15 PACKAGES IN VHDL

3.15.1 Introduction

- VHDL programs model physical systems. There may have some issues, which we have to deal with such as the following issues :
  1. Can *wait* statement be used in procedure ?
  2. Can *signals* be passed to procedures and the modified within the procedure ?
  3. How are *procedures* synthesized ?
  4. Can *functions* operate an signals ?

3.15.2 VHDL Tools

- The tools available in VHDL to solve above issues are as under :
  1. Procedures,   2. Functions,
  3. Packages,   4. Libraries.

3.15.3 Definition

- **A package in VHDL is a collection of functions, procedures, shared variables, constants, files, aliases, types, subtypes, attributes, and components.**

- **The package is a unit that groups various declarations, which can be shared among several designs.**

- **A package is a VHDL file, which can be used to contain user defined data types, constants, functions, procedures etc.** A single package can be shared across many VHDL designs.

- A package is a common storage area used to hold the data to be shared among a number of entities.

- A package declaration can be defined as a set of declaration that may be shared by many VHDL design units.

3.15.4 Description

- A package can be thought of a VHDL tool box that contains several tools to buit VHDL design.

- A package provides a convenient mechanism to store and share declarations that are common across many VHDL design units.

- The purpose of a package is to declare shareable types, subtypes, constants, signals, files, aliases, component, attributes and groups. Once a package is defined, it can be used in multiple independent designs.

- Packages are stored in libraries for greater convenience. A package consists of *package declaration* (mandatory) and may contain a single optimal *package body*.

- Items declared in a package declaration are visible in other design units if the use clause is applied.

- The two-part specification of a package (declaration and body) allows to declare the so-called *deferred constants*, which have no value assigned in the package declaration. The value for a deferred constant, however, must be declared in the package body accompaying the package declaration.

- The VHDL Language Standard defines two standard packages, which must be available in any VHDL environment - package STANDARD and package TEXTIO. The former contains basic declarations of types, constants and operators, while the latter defines operations for manipulating text files. Both are located in the library STD.

- Apart from the VHDL Language Standard there is another standard, which extends the language and supports the extensions in the form of a package, *Std_Logic_1164*.

- Packages in conjunction with libraries are an excellent way that VHDL allows the digital designer to organize his/her VHDL code. By grouping functionality that belongs together, design make much more sense and are leaner. They also allow you to write code that is **reusable**.

- You might find one particular package file being used again and again throughout your designs if it has some useful functions or constants.
- It is preferable to put all of your component definitions in a single package file, rather than copy and pasting them everywhere the component is instantiated. This way, if the port map changes, only the package file and the actual instantiations need to be updated.
- Constants and types that appear repeatedly throughout your code should likely be grouped together in a package file or maybe be passed in your design using generics.
- Functions and procedures need to exist both in the declaration section as well as the body section. As was said before, the declaration contains the prototype for the function or procedure and the body contains the actual implementation of the code.
- In order to use the constants, components, functions, etc that you created in your package file, you will need to tell your other code where to look for these things. This is done with the *use* clause. The end of the use clause tells the file to use everything inside of the package file, as opposed to one particular function

3.15.5 Simplified Syntax

- The simplified syntax for a package in VHDL is as under :

  package package_name is

  package_declarations

  end package package_name;

3.15.6 Uses

1. It is used to keep user defined functions and procedures in a common place.
2. It is used to declare custom data types.

3.16 CONFIGURATION IN VHDL

3.16.1 Introduction

- In case of "plugging chips into sockets", it was suggested in the default binding section that an instantiation of a design entity need not have a component declaration of the same name to be legal VHDL, providing we use a configuration to change the default binding.
- The emphasis was on establishing the relationships between component declaration, design entity (particularly its entity declaration) and component instantiation.
- In entity declaration of configurations, we will look at selecting one architecture from many architectures of one design entity for instantiation essentially specifying which die goes in the package of the chip that will be plugged into the PCB to maintain our analogy, and in architecture body, we will look at choosing from amongst different design entities for instantiation essentially specifying which chip to plug into the socket.
- The configuration is an advanced concept in VHDL, but it can be very useful, when used properly.

3.16.2 Definition

- **A configuration is a construct that defines how component instances in a given block are bound to design entities in order to describe how design entities are put together to form a complete design.**
- **A configuration is a construct that defines how the design hierarchy is linked together.**
- A configuration declaration defines how the design hierarchy is linked together during elaboration, by listening the entities and architectures used in place of each component instantiation within an architecture.
- A declaration is a separate unit, which allows for late binding of components i.e., the binding can be performed after the architecture body has been written.

3.16.3 Description

- The configuration does the job of specifying the exact set of entities and architectures used in a particular simulation or synthesis run.

- A configuration does two things. Firstly, a configuration specifies the design entity used in place of each component instance i.e., it plugs the chip into the chip socket and then the socket-chip assembly into the PCB. Secondly, a configuration specifies the architecture to be used for each design entity i.e., which die.

- A VHDL configuration allows the designer to specify different architectures for a single entity. In other words, the intervals of a design can change, while the interface remains the same.

- Configurations are not configured to get a basic VHDL design running, therefore, unless you have a compelling rearm to use them.

- Each component instantiation refers to some design entity (entity/architecture pair) and the association is specified by a *configuration specification*.

- Component specification appears in the declarative part of the unit, where the instances are used. If for some reasons, however, it is appropriate (or desired) to postpone (or defer) such association until later, *configuration declaration* can be used for specifying such deferred component specifications.

- The configuration declaration starts with the configuration name and then it is associated to a given design entity. Declarative part of the configuration may contain *use clauses*, *attribute* specifications and *group* declarations.

- The main part of the configuration declaration contains so called *block configuration*. It indicates which architecture will be used with the entity specified earlier, as well as which configuration elements will be used in the component instantiation.

- The configuration declaration may contain other blocks' configurations, allowing this way to specify hierarchical structures. Such a configuration can be called *hierarchical*, while a configuration without hierarchy can be called *simple*.

- A simple configuration contains reference to only one architecture body. Hierarchical configurations, on the other hand, allow to nest block configurations. This mechanism allows binding component instantiation statements with the design entities down the hierarchy.

- When the ports and generics in component declaration do not match with their counterparts in entity declaration, so called *binding* indication can be applied.

- This is an explicit notification on how the ports and generics in the entity should be bound to ports and generics of the component instance.

- The *generic map* and *port map* clauses are used for this purpose.

3.16.4 Need

- The configurations is need for any one of the following reasons :
  1. Swapping between different directed test architecture is a good use of configurations.
  2. Bus Functional Models (BFMs) can be swapped out and replaced with the actual RTL code by using configurations.
  3. Testing unusual behavior is useful in simulations for finding storage corner cases in a design.
  4. Swapping out large RTL code for base bones overplay code in the interest of simulation speed is a very handy use for configurations.

3.16.5 Simplified Syntax

- The simplified syntax for configuration in VHDL is as :

configuration configuration_name **of** entity_name **is**

  -- configuration declarations

    for architecture_name

```
        for instance_label:component_name
            use entity library_name.entity_name(arch_name);
        end for;
    -- other for clauses
end for;
end [configuration] [configuration_name];
configuration configuration_name of entity_name is
-- configuration declarations
    for architecture_name
        for instance_label:component_name
            use configuration library_name.config_name;
        end for;
    -- other for clauses
end for;
end [configuration] [configuration_name];
```

3.16.6 Important Notes

1. Configuration assigns one and only one architecture to a given entity.

2. Synthesis tools do generally not support configurations.

3. For a configuration of some design entity, both the entity and the configuration must be declared in the same library.

3.17 VHDL LIBRARY

3.17.1 Introduction

- IEEE 1164 standard is a technical standard published by the IEEE in 1993. It describes the definitions of logic values to be used in electronic design automation for the VHDL.

- You can create a VHDL library to manage and store user created VHDL libraries using TCL commands.

3.17.2 Definition

- A library is a storage facility previously analyzed design units. In practices, this relates mostly to packages.

- A VHDL library is a container, which holds files that define entities, architectures, or packages. You can view and manage VHDL libraries in the libraries panel.

- A library clause defines logical names for design libraries which are used the design units in the hast environment.

3.17.3 Description

- In VHDL, the hardware designers makes the declarations visible via library. A library promotes sharing of promoted design and hides the source code from.

- VHDL libraries allow you to store commonly used packages and entities that you can use in your VHDL files.

- A VHDL library is not stored as a separate file on disk. VHDL libraries and their corresponding sources are referenced in the ISE project file and are passed to the synthesis or simulation tool to be compiled.

- In order to specify a package and to make it visible to the designer, the *library clause*, i.e. making the library visible and *use clause*, i.e. making particular declaration visible must be used.

- There are two predefined libraries, which are used implicitly in every design; STD and WORK. The first of them contains standard packages STANDARD and TEXTIO. The other is a working library, where all user-created and analysed design units are stored. The user-specified packages are stored in the working library WORK.

- A library specified in a library clause of the primary design unit (entity, configuration or package) is visible in each secondary unit (architecture or package body) associated to it.

- Library STD (containing packages STANDARD and TEXTIO) need not to be specified. Both packages are automatically included in every design unit.

- Library clause may contain more than one library names separated by commas.

- Work library is the pointer to the working directory, i.e. where all the design files are kept and the software 'knows', which one that is, as it is set up by the project definition. Thus if the user were to put a .vhd file, containing a package, in the current working directory, the statement referencing that package would be: library work; -- not needed, but OK to include.

- A uses libraries uses Synplicity for the compiler, since it does not impose the restrictions of the compiler, but behaves as the language intends. Since the result of this compilation is an EDIF file, it can however, be exported to complier and from then on used as any internal VHDL source file for simulation and implementation in an Altera FPGA.

3.18 VHDL MODELING STYLES

3.18.1 Introduction

- VHDL is a programming language used to model a digital systems by using different styles of modeling.

- Modeling style means that how we design out digital ICs in electronics. With the help of modeling style, we describe the designing electrons.

- In VHDL, an entity is used to describes a hardware module. An entity, can be described by using entity declaration, architecture, configuration, package declaration and package body.

- Hardware is inherently concurrent which means activities can and will happen in parallel. That is whenever there is an event, all consequences of the event will happen at the same time.

- VHDL allows mixing of descriptions of blocks with drift event styles. Thus is a form of concurrent.

- Even though hardware is concurrent, we need to have the capability to describe design or model with sequential statements similar to the conventional procedure language such as C or PASCAL. VHDL supports sequential statement.

3.18.2 Modeling Styles

- There are three ways or styles of VHDL modeling to describe the function of a VHDL model component as under :

  1. Structural modeling,   2. Data flow modeling,
  3. Behavioral modeling.

3.18.3 Structural Modeling

- In this modeling, an entity is described as a set of interconnected components of the structural style of modeling describes only an interaction of components, viewed as black boxes, without implying any behavior of the components themselves, nor of the entity that they collectively represent.

- In structure modeling, architecture body is composed of two parts namely the declarative part before the keyword begin and the statement part after the keyword begin.

- Structural design is simple to understand and this style is the schematic capture and utilizes simple.

- The top-level design entity architecture describes the interconnection of lower-level design entities. Each lower-level design entity can be described as an interconnection design entities at the next lower-level and so on.

- The structural architecture use only component instantiation statement. This modeling style is most useful and efficient, when a complex system is described as an interconnection of moderately complex design entity to be independently designed and reshifted before being used in the higher-level description.
- The structural modeling is very similar to the *schematic entry*, which is implemented as text instead of graphically.

3.18.4 Dataflow Modeling

- Dataflow style describe a system in terms of how data flows through the digital system. Data dependencies in the description may be those in a typical hardware implementation.
- A dataflow description directly implies a corresponding gate-level implementation and it consists of one or more concurrent signal assignment statements.
- The flow of data through the entity in expressed using concurrent, i.e. parallel signal. The concurrent statements in VHDL are WHEN and GENERATE.
- Beside then, assignments using only operators such as AND, NOT, +, *, sll etc. can also be used to construct VHDL code. A special kind of assignments called BLOCK, can also be employed in this kind of VHDL code.
- Dataflow modeling describes the architecture of the entity under design without describing its component in terms of flow of data from input towards output.
- Dataflow style is nearest to RTL description of the digital circuits. This modeling is concurrent style of modeling in VHDL, i.e., unlike behavioral modeling, the order of statements is not important.
- Dataflow modeling can be used to describe combinational digital circuits. The basic mechanism used is the concurrent assignment, in which a value is assigned to a signal.

3.18.5 Behavioral Modeling

- The main VHDL construct for behavioral modeling in hardware is the process statement, which is one of the concurrent statement of the VHDL.
- The process statement has an associated body of sequential statements; which are executed to compute and schedule output changes every time any input signal on the sensitivity channel of the process statements changes.
- Every modeling language models its subject using certain value, sets, operations on those value sets and defining storage locations to hold the values.
- A behavioral description describes a system's behavior or function in an algorithm fashion.
- Behavioral style is the most abstract style. The description is abstract in the sense that it does not directly imply contains gate-level implementation.
- Behavioral style consists of one or more process statements. Each process statement is single concurrent statement that itself contains one or more sequential statements.
- The behavioral modeling describes how the digital circuit should behave. For these reasons, behavioral modeling is considered highest abstraction levels as compared to dataflow or structural models.
- The behavior of the entity is expressed using sequentially executed code, which is very similar in syntax language like C or PASCAL.
- The behavioral model is widely used in test bench design, since the test bench design does not care about the hardware realization.

Practice Questions

1. What is VHDL ? Give its applications.
2. Why do you need VHDL in digital design ?
3. State the important features of VHDL.

4. What are Pro's and Con's of VHDL ?

5. What are the advantages and disadvantages of VHDL ?

6. Compare VHDL with verilog.

7. What are language elements of VHDL ?

8. List object types in VHDL.

9. Explain following with an example :

    (a) Content, (b) Variable, (c) Signal.

10. Compare variable and signal.

11. What are object and data types ?

12. Write a short note on object types or data types.

13. Define the following data types in VHDL :

    (a) Scalar, (b) Composite, (c) Access, (d) File

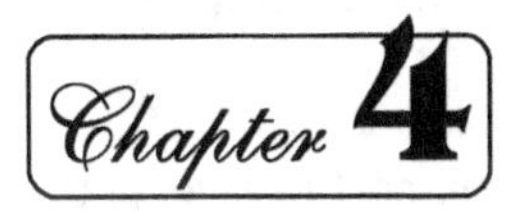

VHDL PROGRAMMING

Syllabus

4.1 Concurrent constructs (when, with).

4.2 Sequential constructs (process, if, case, loop, assert, wait).

4.3 VHDL program to implement flip-flop, Counter, Shift register, MUX, DEMUX, ENCODER, DECODER, MOORE, MEALY machines.

4.4 Test bench and its applications.

Learning Objectives

(4a) Develop program using concurrent statements for the given application in VHDL.

(4b) Develop program using sequential statements for the given application in VHDL.

(4c) Develop program to implement the given combinational/sequential logic circuit using VHDL.

(4d) Describe the test bench for the given application in VHDL.

4.1 STATEMENTS IN VHDL

4.1.1 Introduction

- A VHDL description has two domains, namely a *sequential domain* and a *concurrent domain*.
- The sequential domain is represented by a process or a subprogram that contains sequential statements. These statements are executed in the order in which they appear within the process or subprogram as in programming languages.
- The concurrent domain is represented by an architecture that contains process, concurrent procedure calls, concurrent signal assignment, and component instantiations.

4.1.2 Types

- There are two types of VHDL statements as under :
  1. Sequential statements
  2. Concurrent statements

4.2 SEQUENTIAL (CONSTRAINTS) STATEMENTS

4.2.1 Definition

- **A sequential statement is defined as algorithms, which are used for execution within a process or subprogram.**

4.2.2 Need

- VHDL is hardware description language, which can be used to tell the synthesis software what physical components need to be added to the design and how these components are connected to each other.

4.2.3 Description

- Sequential statements allow use to describe the abstract behavior of a circuit rather than use low-level components, such as different logic gates, to build the circuit. This abstract behavior description can sometimes make the circuit design simpler. This is due to the fact that human reasoning and algorithms, resemble a sequential access.
- Unlike *concurrent statements*, the frequency statements are executed line by line. Hence, we need to separate the two types of code from each other. This is done by enclosing the VHDL construct known as 'process'. A process can appear anywhere after the 'begin' statement of the architecture.

- Sometimes, the use of sequential statements is not only simpler, but also safer and more efficient.
- The basic structure of a process suggests that the sequential statements should be placed between the keywords *'begin'* and *'end'*. In this part the code, the order of the statement is important just like a code in a traditional computer programming.
- Sequential statements define algorithms for the execution within a process or subprogram.
- A process contain sequential statements that describe the behaviour of an architecture. Its process can also be seen as an infinite loop.
- A VHDL process permits concurrent algorithms, i.e. a number of different processes may run at the simulation time.
- Sequential statements like n = 3 : = 3 are interpreted one after the another, in the order in which they are written.
- VHDL sequential statements can appear only in a process or subprogram. A VHDL process is a group of sequential statements; subprogram is a procedure or an function.
- Processes are composed of sequential statements, but processors themselves are concurrent statements. All processes in a design execute concurrently. However, at any given time only one sequential statement is interpreted within each process.
- A process communicates with the rest of a design by reading or writing values to and from signals or parts decaled outside the process. Sequential algorithms can be expressed as sub-programs and called sequentially or concurrently.

4.2.4 Types

- There are different types of commonly used sequential statements.
- They are as under :
  1. Assignment statement
  2. Variable assignment statements,
  3. Signal assignment statement
  4. if statement
  5. Case statement
  6. Loop statement
  7. Next statement
  8. Exit statement
  9. Subprograms
  10. Return statement
  11. Wait statement
  12. Null statements

4.2.5 Assignment Statements

1. Description :

- An assignment statement assigns a value to a variable signal. A syntax is as under :
  target : = expression; -- *Variable assignment*
  target < = expression; -- *Signal assignment*
- Target is a variable or signal or part of a variable of signal, such as a sub array that receives the value of the expression. The expression must be evaluated to the same type as the target.
- The difference in sentence between variable assignments and signal assignments is that variables use : = and signals use =. The basic semantic difference in that variables are local to a process or subprogram, and their assignments take effect immediately.
- Signals need not be liked to a process or subprogram, and their assignment take effect at the end of a process. Signals are the only means of communication between process.

2. Assignment Targets :

- Assignment statements have five kinds of targets :
  (a) Simple names, such as my_var
  (b) Indexed names, such as my_array_var (3)
  (c) Slices, such as my_array_var (3 to 6)
  (d) Field names, such as my_record a field
  (e) Aggregates, such as (my_var 1, my_var 2)
- A assignment target can be either a variable or a signal decription.

3. Simple Name Targets :

- The syntax for an assignment to simple name target is
  identifier : = expression; -- *Variable assignment*
  identifier < = expression; -- *Signal assignment*

- Identifier is the name of a signal of variable. The assigned expression must have the same type as the signal or variable for array types, all elements of the array are assigned variables.
- Assignments to simple name targets is shown below :

```
     Variable A, B  :  BIT;
          Signal C  :  BIT_VECTOR (1 to 4);
          -- target Expression
                    A  :  '1';        -- variable A is assigned '1'
                    B  :  '0';        -- variable B is assigned '0'
                    C  :  '1100';     -- signal array C is assigned '1100'
```

4. Indexed Name Targets :

- The syntax for assignment to an indexed name target is

identifier (index_expansion) : = expression; -- Variable assignment

identifier (index_expression) < = expression; -- Signal assignment

- identifier is the name of an array type signal or variable index_expression must valuate to an index value for the identifier array's index type and bounds. It does not have to be comparable, but more hardware is synthesized, if it is not.
- The assigned expression must have the array's element type. Array variable elements are assigned values or indexed names.

```
     Variable A, B  :  BIT_VECTOR (1 to 4);
          -- Target      Expression
                    A (1)  :  '1';    -- Assigns '1' to the first element of array A.
                    A (2)  :  '1';    -- Assigns '1' to the second element of array A.
                    A (3)  :  '0';    -- Assigns '0' to the third element of array A.
                    A (4)  :  '0';    -- Assigns '0' to the fourth element of array A.
```

- The above example shows two indexed name targets. One is computable and the other is not. Note the two differences in the hardware generated for each assignment.

5. Slice Targets :

- The syntax for slice target is as under :

identifier (index_expr_1 direction index_expr_2)

- identifier is the name of an array type signal or variable. Each index_expr expression must evaluate to an index value for the identifier array's index type and bounds.
- Both index_expr expressions must be computable and must lie within the bounds of the array. Direction must match the identifier array type's direction either to or down to.
- The assigned expression must have the array's element type. Array variables A and B are assigned the same value.

```
     Variable A, B  :  BIT_VECTOR (1 to 4);
          -- Target      Expression
               A (1 to 2)  :  "11";   -- Assigns "11" to the first two elements of array A.
               A (3 to 4)  :  "00";   -- Assigns "00" to the last two elements of array A.
               A (1 to 4)  :  "1100"; -- Assigns "1100" to array B.
```

6. Field Targets :

- The syntax for field target is as under :

identifier.field_name

- identifier is the name of a record type signal or variable and field_name is the name of a field in that record type, preceded by a period (·).
- The assigned expression must have the identified field type. A field can be of any type, including an array, record or aggregate type.
- Assign value to the fields of record variables A and B.

```
type REC is
    record
        NUM_FIELD : INTEGER range - 16 to 15;
        ARRAY_FIELD : BIT_VECTOR (3 to 0);
    end record;
variable A, B : REC;
        --Target                 Expressions;
          A.NUM_FIELD :  = - 12;         -- Assigns -12 to record A's
                                         -- field NUM_FIELD
          A.ARRAY_FIELD :  = "1100";     -- Assigns "1100" to record A's
                                         -- field ARRAY_FIELD
        A.ARRAY_FIELD (3) :  = '1';      -- Assigns '1' to the MSB of record
                                         -- A's field ARRAY_FIELD
                          :  = A;        -- Assigns values of record
                                         -- A to corresponding fields of B.
```

7. **Aggregate Targets :**
*   The syntax for assignment to an aggregate target is
    ([choice=>] identifier
    {, [choice =>] identifier}) : = array_expression;
    -- variable assignment
    ([choice=>] identifier
    {, [choice =>] identifier}) : <= array_expression;
    -- signal assignment
*   An aggregate assignment assigns array_expression's element values to one or more variable or signal identifiers.
*   Each (optional) *choice* is an index expression selecting an element or a slice of the assigned array array_expresssion. Each identifier must have array_expression's element type. An *identifier* can be an array type. Some aggregate targets are shown below :

```
signal A, B, C, D : BIT;
signal S : BIT_VECTOR (1 to 4);
variable E, F : BIT;
variable G : BIT_VECTOR (1 to 2);
variable H : BIT_VECTOR (1 to 4);
    -- positional notation
    S < = ('0', '1', '0', '0');
    (A, B, C, D) < = S ;            -- Assigns "0" to A
                                    -- Assigns "1" to B
                                    -- Assigns "0" to C
                                    -- Assigns "0" to D
    --- Named notation
    (3) = > E, 4 = > F,
    2 = > G(1), 1 = > G(2) :  = H;  -- Assigns H(1) to G(2)
                                    -- Assigns H(2) to G(1)
                                    -- Assigns H(3) to E
                                    -- Assigns H(4) to F
```

*   You can assign array element values to the *identifiers* by position or by *name*. In positional notation, the *choice* = > construct is not used. *Identifiers* are assigned array element values in order, from the left array bound to the right array bound.

- In named notation, the *choice – >* construct identifies specific elements of the assigned array. A choice expression index indicates a single element such as 3. The *identifiers* type must match the assigned expression's element type.
- Positional and named notations can be mixed, but positional associations must came before named associations.

4.2.6 Variable Assignment Statements

- A variable assignment changes the value of a variable. The syntax is

target ; = expression;

- **expression** determines the assigned value. Its type must be compatible with target. *target* names the variables that receive the value of expression.
- When a variable is assigned a value, the assignment takes place immediately. A variable keeps its assigned value until another assignment.

4.2.7 Signal Assignment Statement

- A signal assignment changes the value being driven on a signal by the current process. The syntax is

target < = expression;

- expression determines the assigned value. Its type must be compatible with target. target names the signal that receive the value of expression.
- Signals and variables behave differently when they are assigned values. The difference lie in the way the two kinds of assignments take effect and how that affects the values read from either variables or signals.

4.2.8 Case Statement

- The case statement selects for execution of one of several alternative sequences of statements, the alternative is chosen based on the value of the associated expression. Thus, the case statement executes one of several sequences of statements, depending on the values of a single expression. The simplified syntax is as under :

case expression **is**
    when choices = > {sequential_statement}
    (**when** choices = > {sequential_statement})
end_case;

- Each choice can be either a static expression such as 3 or a static range such as 1 to 3. The type of *choice_expression* determines the type of each *choice*. Each value in the range of *choice_expression* type must be covered by one *choice*.
- The final *choice* can be others, which matches all remaining unchosen values in the range of *expression's* type. The others *choice*, if present, matches *expression* only if no other choices match.
- The *case* statement evaluates expression and compares that value to each *choice* value. The *when* clause with the matching *choice* value has its statement executed.
- The following restrictions are placed on choices :
  (1) No two choices can overlap.
  (2) If no *other's* choice is present, all possible values of *expression* must be converted by the set of choices.

4.2.9 Loop Statement

- **Statement that includes a sequence of statements that is to be executed repeatedly, zero or more times is called a loop statement.**
- Loops operate in the usual way, i.e. they are used to execute the same VHDL code a couple of times.
- The loop variable is the only object in VHDL, which is implicitly defined. The loop variables cannot be declared externally and is only visible within the loop. Its value is only read, i.e. the number of cycles is fixed, when the execution of the *'for loop'* begins.
- The loop statement with no iteration scheme repeats enclosed statements forever. The simplified syntax is   [label :] **Loop**
      {sequential_statement}
    end loop [label];

- The optional *label* names this loop *sequential_statement* can be any statement. Two sequential statements are used only with loops, namely the *next* statement, which skips the remainder of the current loop interaction and the exit statement, which terminates the loop.
- A loop statements must have at least one **wait** statement in each enclosed logic branch.

4.2.10 While . . . loop Statement

- The *while ... loop* statement repeats enclosed statements as long as its iteration condition evaluates true. The syntax is as under :

[label ;] I **while** condition loop
     {sequential_statement}
end loop [label];

- The optional *label* names this loop. *condition* is any Boolean expression such as (A = '1') or (X = 0).
- *sequential_statement* can be any statement. Two sequential statements are used only with loop, namely the *next* statement, which skips the remainder of the current loop iteration and the **exit** statement, which terminates the loop.
- A while ... loop statement must have at least one *wait* statement in each enclosed logic branch.

4.2.11 for . . . loop Statement

- The *for ... loop* statement repeats enclosed statement one for each value in integer range. The syntax is

[label :] **for** identifier **in** range loop
     {sequential_statement}
end loop [label];

- The optional *label* names this loop. The user of *identifier* is specific to the for ... loop statement.
  1. identifier is not declared else where. It is automatically declared by the loop itself, and is local to the loop. A loop identifier over rides any other identifier with the same name but only within the loop.
  2. identifier's value can be read only inside its loop. It does not exist outside the loop, you cannot assign a value to a loop identifier.
- VHDL compiler currently requires that range must be a computable integer range in either of two forms as under :

integer_expression **to** integer_expression

integer_expression **downto** integer_expression

- Each integer_expression evaluates to an integer. *sequential_statement* can be any statement. Two sequential statements are used only with loops, namely the next statement, which skips the remainder of the current loop iteration and the *exit* statement, which terminates the loop.
- A for....loop statement must not contain any wait statements. A for....loop statement executes as :
  1. A new local, integer variable is declared with the name **identifier**.
  2. *identifier* is assigned the first value of **range**, and the sequence of statements is executed once.
  3. *identifier* is assigned the next value in *range*, and the sequence of statements is executed once more.
  4. Step (3) is repeated unit *identifier* is assigned to the last value in range. The sequence of statements is then executed for the last time, and execution continues with the statement end loop. The loop is then inaccessible.
- Two equivalent code fragments of *for...loop* statements are shown below :

variable A, B: BIT_VECTOR(1 to 3);
-- First fragment is a loop statement for I in 1 to 3 loop
     A(I) <= B(I);
end loop;
-- Second fragment is three equivalent statements
     A(1) <= B(1);
     A(2) <= B(2);
     A(3) <= B(3);

- You can use a loop statement to operate on all elements of an array, without explicitly depending on the size of the array. The for...*loop* statement operating on an *Entire Array* is shown below :

```
variable A, B: BIT_VECTOR(1 to 10);

. . .

for I in A' range loop
    A(I) := not B(I);
end loop;
```

4.3 CONCURRENT STATEMENTS

4.3.1 Definition

- **A concurrent statement in VHDL is a signal assignment within the architecture, but outside of a normal process construct.**
- VHDL permits concurrent algorithms, i.e. a number of different processes may run at the same simulation time.
- Concurrent statements are used to define interconnected blocks and processes that jointly describe the overall behavior or structure of a design.
- *The primary concurrent statements are the block statements, which group together other concurrent statements and the process statement, which represents a single independent sequential process.*

4.3.2 Description

- Concurrent statement execute asynchronously with respect to each other.
- The processes communicate with each other through signals. A process contains sequential statement that describes the behavior of an architecture. A process also can be seen as an infinite loop. The processes must contain either an explicit sensitivity list as a *wait* statement.
- Additional concurrent statements provide convenient syntax for representing simple, commonly occurring forms of processes, as well as for representing structural decomposition and regular description.
- Within a given simulation cycle, an implementation may execute concurrent statement in parallel or in some order. The language does not define the order, if any, in which such statements will be executed. A description that depends upon a particular order of execution of concurrent statements is erroneous.
- All concurrent statements may be labelled, such labels are implicitly declared at the beginning of the declarative part of the innermost enclosing entity declaration, architecture body, block statement, or generate statement.

4.3.3 Types

- There are several related concurrent statements equivalent to the process statement. These statements can be organized into blocks to provide hierarchy or to improve readability. The variable VHDL concurrent statements are as follows :

  1. Process statement
  2. Concurrent signal assignment statement
  3. Conditional signal assignment statement
  4. Selective signal assignment statement
  5. Concurrent procedure call statement
  6. Block statement
  7. Concurrent assertion statement
  8. Concurrent instantiation statement
  9. Generate statement

4.3.4 Process Statement

- The process statement is very similar to the classical programming language. The code inside the process statement is executed sequentially.
- The process statement is declared in the concurrent section of the architecture, so two different processes are executed concurrently.
- The process statement encloses a set of sequentially executed statements. Statements within the process are executed in the order they are written. However, when viewed from outside, a process statement is a single concurrent statement.

- The content of the process statement can include sequential statement, like those found in software programming languages. The program statements are used to compute the output of the process from its inputs.
- The VHDL process syntax contains sensitivity list, declarative list and sequential statement section. The declaration process statement is as under :

  process_label : **process**(sensitivity_list)
  -- declarative **part**
  begin
  -- sequential statement
  end process process_label;

- The process label is optional, you can avoid using label. Labelling all processes you use, the code will be clear and it will be simple to arrange the simulation environment.
- The process sensitivity lists the signal that will cause the process statement to be executed.
- Any transaction on any of the signal is the sensitivity list which cause the process to execute.
- Signal talents in the sensitivity list are as under :
  1. Signals on the right hand side of a assignment statement.
  2. Signals used in conditional expressions.
- The process statements are used to compute the outputs of the process from its used.

4.3.5 Conditional Statements

- When we read to perform a choice or selection between two or more choices, we can use the VHDL conditional statement.
- We can use a conditional expression with the *wait* until statement. The expression ensures that the process is triggered when the two motor signals are equal.
- The *if_then_elsif* statements can be used to create branches in the program. Depending on the value of a variable, or the outcome of an expression the program can take different paths. The basic syntax is :

  if <condition then
       elsif <condition then
  else
  end if ;

- The *elsif* and *else* are optional, and *elsif* may be used multiple times. The <condition> can be a boolean true or false, or it can be an expression, which evaluate to true or false.
- The relational operators are as under :
  1. equal =   2. not equal ≠
  3. less than <   4. less than or equal < =
  5. greater than >   6. greater than or equal > =
- The logical operators are as under :
  1. true if a is false – **not** a   2. true if a and b are true – a **and** b
  3. true if a or b are true – a **or** b,   4. true if a or b is false – a **nand** b,
  5. true if a and b are false – a **nor** b,   6. true if exactly one of a or b is true – a **xor** b,
  7. true if a and b are equal – a **xnor** b
- Since the VHDL is a concurrent language it provides two different solutions to implement a conditional statement as under :
  1. Sequential conditional statement
  2. Concurrent conditional statement

4.3.6 Sequential Conditional Statement

- The sequential conditional statement can be used in *process* and *subprogram*. The simplified syntax for a sequential conditional statement is as under :

1. if_statement ::=
    if condition then
        sequence_of_statements
        { elsif condition then
    sequence_of_statements }
  [else
    sequence_of_statements
  ]
  end if ;

2. if boolean_expr_1 then
    sequential_statements;
  elsif boolean_expr_2 then
    sequential_statements ;
  elsif boolean_expr_3 then
    sequential_statements;
  else
    sequential_statements;
  end if;

4.3.7 Concurrent Conditional Statement

- The concurrent conditional statement can be used in the architecture concurrent section, i.e. between the *"begin-end"* section of the VHDL architecture definition. The simplified syntax for concurrent conditional statement is as under :

conditional_signal_assignment ::= target <= conditional_waveforms ;
conditional_waveforms ::=
    { waveform when
  condition else } waveform

4.3.8 Concurrent Signal Assignment Statements

- A concurrent signal assignment statement is equivalent to a process containing only that statement such a statement is expressed in parallel with other concurrent statements are other processes. There are three types of concurrent signal assignment statements as under :

1. Simple signal assignment statement
2. Conditional signal assignment statement
3. Selected signal assignment statement

1. Simple Signal Assignment Statement :

- This statement is the concurrent version of the sequential signal assignment statement and has the same form with this. As the sequential version, the concurrent assignment defines a new driver for the assigned signal.
- A concurrent assignment statement appears outside a process, within an architecture. A concurrent assignment statement represents a simplified form of writing a process and it is equivalent to a process that contains a single sequential assignment statement.
- The syntax for to adder can be written as below :

entity add_1 is
    port (a, b, cin: in bit;
        s, cout: out bit);
  end add_1;
architecture concurrent of add_1 is
    signal S1, S2, S3, S4: bit;

```
begin
    S1 <= b xor cin;
    S2 <= a and b;
    S3 <= a and cin;
    S4 <= b and cin;
    S <= a xor S1;
    cout <= S2 or S3 or S4;
end concurrent;
```

- The concurrent assignment statements appear directly in the architecture, not inside a process. The order in which the statements are written is irrelevant. At simulation all statements are executed in the same simulation cycle. In the case of processes, their activation is determined by the change of a signal in their sensitivity lists or by encountering a wait statement. In the case of concurrent assignment statements, the change of any signal that appears in the right-hand side of the assignment symbol activates the assignment execution, without explicitly specifying a sensitivity list.
- Activation of an assignment statement is independent of activation of other concurrent statements within the architecture.
- The concurrent assignment statements are used for dataflow descriptions. By synthesizing these statements, combinational circuits are obtained.
- If there are several concurrent assignments to the same signal in an architecture, multiple drivers will be created for that signal. In these cases, either a predefined resolution function or a resolution function defined by the user must exist for the signal type.
- As opposed to concurrent assignments, if a process contains several sequential assignments to the same signal, only the last assignment will be effective.

2. Conditional Signal Assignment Statement :

- The conditional assignment statement is functionally equivalent to the *if* conditional statement and has the following syntax:

```
signal <= [expression when condition else ...]
          expression;
```

- The value of one of the source expressions is assigned to the target signal. The expression assigned will be the first one whose associated Boolean condition is true.
- When executing a conditional assignment statement, the conditions are tested in the order in which they are written.
- When the first condition that evaluates to value true is encountered, its corresponding expression is assigned to the target signal.
- If none of the conditions evaluates to value true, then expression corresponding to the else clause is assigned to the target signal.

Differences :

- The difference between the conditional assignment statement and the condition *if* statement are as follows :
  1. The conditional assignment statement is a concurrent statement, and therefore it can be used in an architecture, while the *if* statement is a sequential statement and can be used only inside a process.
  2. The conditional assignment statement can only be used to assign values to signals, while the *if* statement can be used to execute any sequential statement.
- We can define an entity and two architectures for a two-input XOR gate. The first architecture uses a conditional assignment statement, while the second uses an equivalent *if* statement.

```
entity xor2 is
    part (a, b: in bit;
        x: out bit);
```

```
    end xor2;
    architecture arch1_xor2 of xor2 is
    begin
        x <= '0' when a = b else
             '1';
    end arch1_xor2;
    architecture arch2_xor2 of xor2 is
    begin
        process (a, b)
        begin
            if a = b then x <= '0';
                     else x <= '1';
            end if;
        end process;
    end arch2_xor2;
```

4.3.9 Selected Signal Assignment Statement

- Like the conditional signal assignment statement, the selected signal assignment statement allows to select a source expressions based on a condition.
- The difference is that the selected signal assignment statement uses a single condition to select between several options. This statement is functionally equivalent to the *case* sequential statement. The syntax is the following:

with selection_expression **select**
    signal <= expression_1 **when** options_1,

...

expression_n **when** options_n,
[expression **when others**];

- The target signal is assigned the value of one of the expressions. The selected expression is the first from those expressions whose options include the value of the selection expression.
- The syntax of the options is the same as for the *case* statement. Thus, each option may be represented by an individual value or a set of values.
- If an option is represented by a set of values, either the individual values of the set may be specified, separated by the "|" symbol, or the range of values, or a combination of these. The type of the selection expression determines the type of each option.
- All values from the selection expression's range must be covered by an option. The last option may be indicated by the others keyword, which specifies all the values from the selection expression's range that were not covered by the preceding options. The following constraints exist for the various options:
  1. The values from the options cannot overlap each other.
  2. If the others option is missing, all the possible values of the selection expression must be covered by the set of options.
- Note that the options in the selected signal assignment statement are separated by commas.
- The two-input XOR gate definition is modified to use a selected signal assignment statement. The equivalent form of this architecture using a **case** statement has presented as below :

```
entity xor2 is
    part (a, b: in bit;
        x: out bit);
end xor2;
architecture arch_xor2 of xor2 is
```

```
    signal temp: bit_vector (1 downto 0);
begin
    temp <= a & b;
    with temp select
    x <= '0' when "00",
    x <= '1' when "01",
    x <= '1' when "10",
    x <= '0' when "11";
end arch_xor2;
```

4.3.10 Selected Signal Assignment Statement

- This statement is the concurrent version of the sequential **assert** statement, with the same syntax as its sequential version:

  assert condition

      [**report** character_string]

      [**sensitivity** severity_level];

- The concurrent *assert* statement is executed whenever one of the signals in the conditional expression changes, as opposed to the sequential *assert* statement, which is executed when this statement is reached in a process or subprogram.

4.4 VHDL PROGRAMMING FOR DIGITAL CIRCUITS

4.4.1 Implementation of Flip-Flops

(A) D-type Flip-Flop :

- Model can be triggered on the rising or the falling edge of the scalar input port CLK.

- The polarity of CLK is controlled by the parameter CLK_TYPE (Rising, Falling).

- In Asynchronous mode, the Enable input has a higher priority over the CLK input.

- In Synchronous mode, the CLK input has a higher priority over the Enable input.

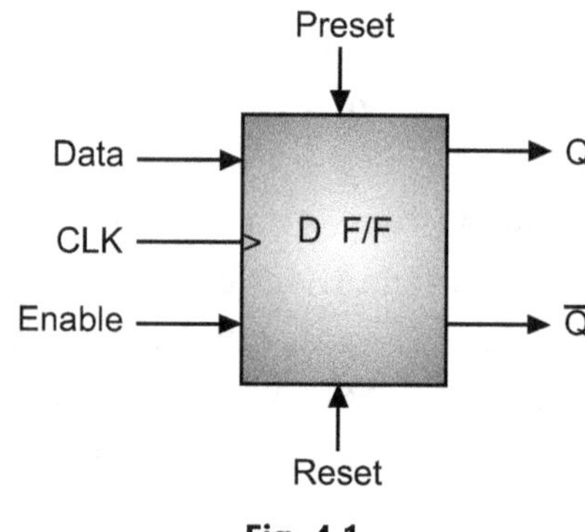

Fig. 4.1

(i) Rising (Positive) Edge Flip-Flop Using IF Statement:

- It is behavioral model VHDL code.

- Rising edge of the CLK is specified by the condition "CLK" event and CLK = "1".

- Process sensitivity list contains only the CLK signal, because it is the only signal that can cause a change in the Q output.

```
Library IEEE;
Use IEEE.Std_logic_1164.all;
entity DFF is
    port (Data, CLK : in Std_logic;
        Q : out std_logic);
end DFF;
architecture Behav of DFF is
begin
    process (CLK)
```

```
        begin
            if (CLK'event and CLK = '1') then
            Q <= Data;
        end if;
    end process;
end behav;
```

(ii) Negative Edge-triggered D Flip Flop:

- Shows behavioral model VHDL code.
- Q follows Data only at the falling edge of the CLK and it is specified by the condition "CLK" event and CLK = '0'."

```
Library IEEE;
Use IEEE.Std_logic_1164.all;
entity DFF_NEG is
    port (Data, CLK : in Std_logic;
            Q : out std_logic);
end DFF_NEG;
architecture RTL of DFF_NEG is
begin
    process (CLK)
        begin
            if (CLK'event and CLK = '0') then
            Q <= Data;
        end if;
    end process;
end RTL;
```

(iii) Rising Edge Flip-Flop with Asynchronous Reset and Clock Enable:

```
Library IEEE;
Use IEEE.Std_logic_1164.all;
entity DFF_CLK_EN is
    port (Data, CLK, Reset, EN : in Std_logic;
            Q : out std_logic);
end DFF_CLK_EN;
architecture behav of DFF_CLK_EN is
begin
    process (CLK, Reset)
        begin
            if (Reset = '0') then
                Q <= '0';
            elsif (CLK'event and CLK = '1') then
                if (EN = '1') then
                Q <= Data;
                end if;
        end if;
    end process;
end behav;
```

(iv) J-K Flip-Flop:

Table 4.1: Truth Table of J-K F/F

J	K	CLK	Q_{n+1}
0	0	↑	Q_n
0	1	↑	0
1	0	↑	1
1	1	↑	$\overline{Q_n}$
×	×	↓	Q_n

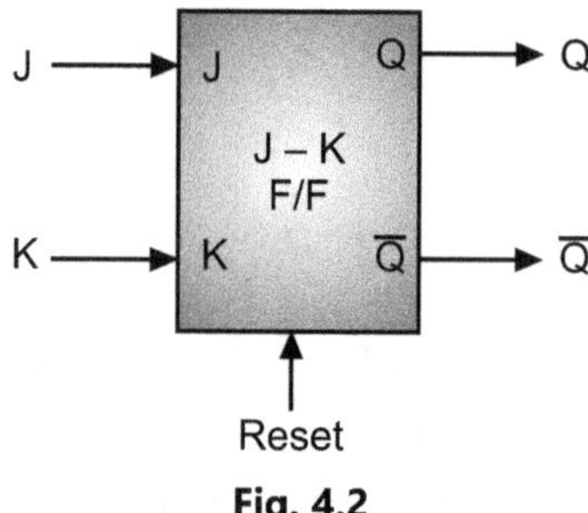

Fig. 4.2

- When you infer a JK flip-flop, make sure you can control the J-K, and CLK signals from the top-level design ports to ensure that simulation can initialize the design.
- The J and K signals act as active high synchronous Set and Reset.

```vhdl
Library IEEE;
Use IEEE.Std_logic_1164.all;
entity JK_FF is
    Port (J, K, CLK, Reset : in Std_ulogic;
        Q, Qbar : out Std_ulogic);
end JK_FF;
architecture JKS of JK-FF is
    signal state : std_ulogic;
begin
    process (CLK, Reset) is
        begin
            if (Reset = '0') then
                State <= '0';
            elsif rising_edge (CLK) then
            case Std_ulogic_vector' (J, K) is
                When "11" =>
                    State <= not state;
                When "10" =>
                    State <= '1';
                When "01" =>
                    State <= '0';
                When others =>
                    null;
            end case;
        end if;
    end process;
    Q <= state;
    Q bar <= not state;
end JKS;
```

4.4.2 Implementation of Shift Registers

- Main use of Shift Register is for converting from serial data input to a parallel data output or vice versa.
- For serial to parallel data conversion, the bits are shifted into the register at each clock cycle, and when all the bits (i.e. 8 bits) are shifted in, the 8-bit register can be read to produce the 8-bit parallel output.
- For a parallel to serial conversion, the 8-bit register first loaded with the input data. The bits are then individually shifted out, one bit per clock cycle, on the serial output line.
- A shift register is characterized by the following control and data signals :
  - Clock
  - Serial input
  - Asynchronous Set/Reset
  - Synchronous Set/Reset
  - Synchronous/Asynchronous parallel load
  - Clock enable
  - Serial or parallel output
  - Shift modes : Left, right etc.

1. **8-bit shift-left register with positive-edge CLK Serial In and Serial Out:**

```
Library IEEE;
Use IEEE.Std_Logic_1164.all;
entity shift_siso is
    port (CLK, Sin : in std_logic);
        Sout : out std_logic);
end shift_siso;
architecture behav of shift_siso is
Signal Temp : Std_logic_Vector (7 downto 0);
    begin
        process (CLK)
            begin
                if (CLK'event and CLK = '1') then
                    for i in 0 to 6 loop
                        Temp (i+1) <= Temp (i);
                    end loop;
                        Temp (0) <= Sin;
                    end if;
                end process;
            Sout <= Temp(7);
            end behav;
```

2. **8-bit shift-left register with positive-edge CLK Serial In and Parallel Out :**

- 8-data bits are shifted in by 8-CLK pulses via a single wire at data-in, below, the data becomes available simultaneously on the eight outputs P_{out} (0) to P_{out} (7) after the 8th CLK pulse.

```
Library IEEE;
Use IEEE.Std_logic_1164.all;
entity shift_SIPO is
    port (CLK, Sin : in Std_logic);
        Pout : Out std_logic_vector (7 downto 0));
    end shfit_SIPO;
    architecture exam of shift_SIPO is
    Signal temp : Std_logic_Vector (7 downto 0));
```

```vhdl
begin
    process (CLK)
        begin
            if (CLK'event and CLK = '1') then
                temp <= temp (6 downto 0) and Sin;
            end if;
        end process;
    Pout <= temp;
end exam;
```

3. **4-bit shift-left/shift-right register with positive-edge CLK – Serial In and Parallel Out (SIPO Register) :**

- When mode selector left-right signal is **'0'** then register work as **shift-left** register.
- When mode selector left-right signal is **'1'** then register work as **shift-right** register.

```vhdl
Library IEEE;
Use IEEE.std_logic_1164.all;
entity shift_LR is
    port (CLK, Sin, left_right : in Std_logic);
        Pout : out std_logic_vector (3 downto 0));
    end shfit_LR;
    architecture exam of shift_LR is
    signal temp : Std_logic_vector (3 downto 0));
        begin
            process (CLK)
                begin
                    if (CLK'event and CLK = '1') then
                        if (left_right = '0') then
                            temp <= temp (2 downto 0) and Sin;
                        else
                            temp <= Sin and temp (3 downto 1);
                        end if;
                    end if;
                end process;
            Pout <= temp;
        end exam;
```

4.4.3 Implementation of Counters

1. Binary Up-counter:

- An n-bit binary counter can be constructed using a modified n-bit register, where the data inputs for the register come from an incrementer (adder) for an up-counter. Starting with a value stored in a register, to get to the next up count sequence, we simply have to add a 1 to it.
- A 4 bit binary up-counter with Asynchronous clear.
- The statement use IEEE.Std_logic_Unsigned.all is needed in order to perform additions on Std_logic_vectors.

- The internal signal **value** is used to store the current count. When clear is asserted, value is assigned the value "0000" using the expression OTHERS $\Rightarrow$ '0'.

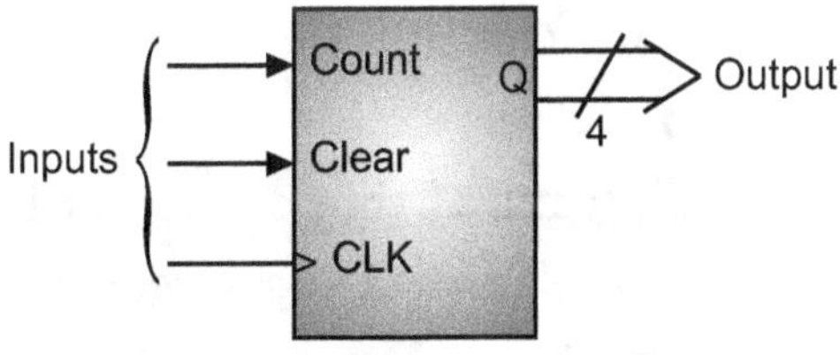

Fig. 4.3 : 4-bit-binary-up-counter diagram

- If count is asserted, the value will be incremented by 1 on the next rising CLK edge.

Library ieee;
Use ieee.std_logic_1164.**all;**
Use ieee.std_logic_unsigned.all
entity counter **is**
    port   (CLK : **in** std_logic;
           Clear : **in** std_logic;
           Count : **in** std_logic;
           Q : out std_logic_vector (3 downto 0)); – – (number of bits can be changed in this line)
end counter;
architecture **behav** of counter is
Signal value:std_logic_vector (3 downto 0); —(number of bits can be changed in this line)
    begin
        Process (CLK, Clear)
           begin
             if Clear = '1' then
               Value <= (Others => '0'); – – (4-bit vector of 0, same as "0000")
             elsif (CLK' event and CLK = '1') then
               if Count = '1' then
                 Value <= Value + 1;
               end if;
             end if;
           end process;
        Q <= Value;
    end behave;

2. Binary Down Counter :

- An n-bit binary counter can be constructed using modified n-bit register where the data inputs for the register come from a decrementer (subtractor) for a down counter.

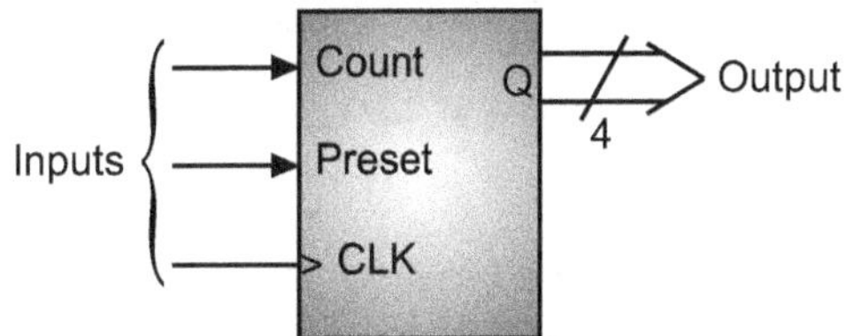

Fig. 4.4 : 4-bit binary down counter

- Starting with a value stored in a register, to get to the next down count sequence, we simply have to subtract a 1 to it.

    Library ieee;
    Use ieee.std_logic_1164.all;
    Use ieee.std_logic_Unsigned.all – – need this to subtract std_logic vectors
    entity counter **is**
        port (CLK : in Std_logic;
           preset : in Std_logic;
           count : in Std_logic;

```vhdl
        Q : Out Std_logic_vector (3 downto 0));
    end counter;
    architecture behav of counter is
    signal value : Std_logic_vector (3 downto 0);
        begin
            process (CLK, preset)
                begin
                    if preset = '1' then
                        value <= (Others => '1'); -- (4-bit vector of 1, same as "1111")
                    elsif (CLK'event and CLK = '1') then
                        if count = '1' then
                            value <= value -1;
                        end if;
                    end if;
                end process;
            Q <= value;
        end behav;
```

3. Binary Up-Down Counter:

- Design of an n-bit up-down counter is just like the up-counter except that we need both an adder and a subtractor for the data input to the register.

- Similar to the up-counter code with the additional logic for the down signal.

- If down is asserted, then value is decremented by 1, otherwise it is incremented by 1.

- For 4-bit binary value, the counter output signal Q is declared as an integer that ranges from 0 to 16. Furthermore, the storage for the current count, value is declared as a variable type integer.

```vhdl
        Library ieee;
        Use ieee.std_logic_1164.all;
        entity UDCOUNTER is
            port (CLK : in Std_logic;
                clear : in Std_logic;
                count : in Std_logic;
                down : in Std_logic;
                Q : Out integer range (0 to 15);
        end UDCOUNTER;
        architecture behav of UPCOUNTER is
            begin
                process (CLK, clear)
                    variable value : integer range 0 to 15;
                    begin
                        if clear = '1' then
                            value := 0
                        elsif (CLK'event and CLK = '1') then
                            if count = '1' then
```

```
                    if down = '0' then
                            value <= value +1;
                    else
                            value <= value -1;
                    end if;
                end if;
            end if;
        Q <= value;
    end process;
end behav;
```

4. 4-bit Synchronous Counter:

```
    Library ieee;
    Use ieee.std_logic_1164.all;
    Use ieee.std_logic_Unsigned.all;
    entity syncontr is
        generic (n:positive :=4);
                        -- Using this statement counter can be programmed for any number of bits.
            port (CLK, reset, enable : in std_logic;
                count : out std_logic_vector ((n-1)downto 0));
        end syncontr;
        architecture CSR of syncontr is
        signal count_int : std_logic_vector((n-1)down to 0);
        begin
            process
                begin
                    wait until rising_edge ((CLK);
                        if reset = '1' then
                            count_int <=(other => '0');
                        elsif enable = '1' then
                            count_int <= count_int +1;
                        else
                            null;
                        end if;
                end process;
            count <= count_int;
        end CSR;
```

4.4.4 Implementation of 4 × 1 MUX

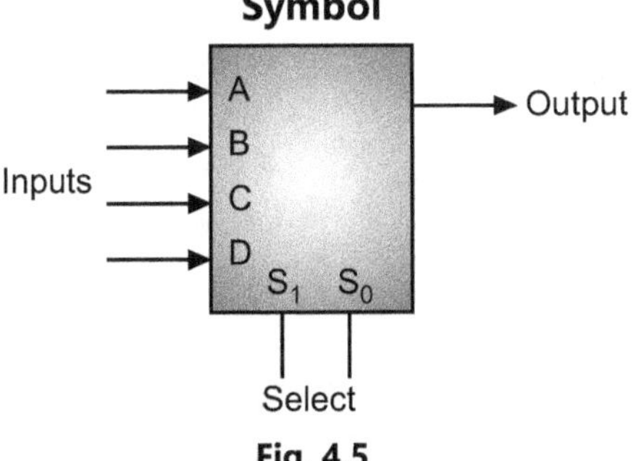

Fig. 4.5

Table 4.2 : Truth table

Select		Output
S_1	S_0	Y
0	0	A
0	1	B
1	0	C
1	1	D

1.   4 × 1 MUX using if statement:

- An **if statement** generally produces **priority-encoded** logic.

```
Library ieee;

Use ieee.std_logic_1164.all;

entity MUX4×1 is

    port (Select : in std_logic_vector (1 downto 0);

        A, B, C, D : in std_logic;

        Y : Out std_logic);

    end MUX4×1;

architecture behav of MUX4×1 is

    begin

        process (Select, A, B, C, D)

            begin

                if (Select = '00') then y <= A;

                elsif (Select = '01') then y <= B;

                elsif (Select = '10') then y <= C;

                else y <=D;

            end if;

        end process;

    end behav;
```

4.4.5 Implementation of DeMUX

1.   1:8 DeMUX using when-else statement:

```
Library ieee;

Use ieee.std_logic_1164.all;

entity demux 1_8 is

    port (din : in std.logic;

        Select : in std_logic_vector (2 downto 0);

        dout : out std_logic_vector (7 downto 0);

    end demux 1-8;

architecture demultiplexer 1_8 of demux 1_8 is

begin

    dout <= (din & "0000000") when (select = "000")
```

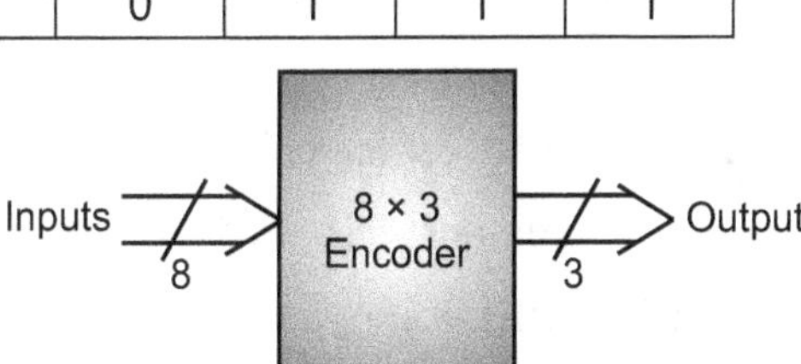

else ('0' & d_{in} & "000000") **when** (select = "001")

else ("00" & d_{in} & "00000") **when** (select = "010")

else ('000' & d_{in} & "0000") **when** (select = "011")

else ('0000' & d_{in} & "000") **when** (select = "100")

else ('00000' & d_{in} & "00") **when** (select = "101")

else ('000000' & d_{in} & "0") **when** (select = "110")

else ('0000000' & d_{in});

    end demultiplexer 1_8;

Fig. 4.6

4.4.6 Implementation of Encoders

- Encoders are used to encode discrete data into a coded form and decoders are used to convert it back into the original undecoded from.

- An encoder having 2^n (or less) input lines encodes input data to provide n encoded output lines.

  e.g. 8 × 3 Encoder.

Table 4.3 : Truth Table

Inputs								Outputs		
I_7	I_6	I_5	I_4	I_3	I_2	I_1	I_0	y_2	y_1	y_0
0	0	0	0	0	0	0	1	0	0	0
0	0	0	0	0	0	1	0	0	0	1
0	0	0	0	0	1	0	0	0	1	0
0	0	0	0	1	0	0	0	0	1	1
0	0	0	1	0	0	0	0	1	0	0
0	0	1	0	0	0	0	0	1	0	1
0	1	0	0	0	0	0	0	1	1	0
1	0	0	0	0	0	0	0	1	1	1

Library ieee;

Use ieee.std_logic_1164.all;

Use ieee.numeric_std.all;

Fig. 4.7

entity Encoder8 is

    port (I : **in** std_logic_vector (7 **downto** 0);

        y : **out** std_logic_vector (2 **downto** 0);

    end Encoder 8;

architecture ARCH of Encoder8 is

    begin

        process (I)

            begin

                Case I is

                    When "00000001" => y <= "000";

                    When "00000010" => y <= "001";

```
            When "00000100" => y <= "010";

            When "00001000" => y <= "011";

            When "00010000" => y <= "100";

            When "00100000" => y <= "101";

            When "01000000" => y <= "110";

            When "10000000" => y <= "111";

            When others => y <= "xxx";

        end case;

    end process;
```

4.4.7 Implementation of Decoder

- Decoders are used to decode data that has been previously encoded using a binary coded format (or other coded format).

- An n-bit code can represent upto 2^n distinct bits of coded information.

- Decoder with n inputs can decode upto 2^n outputs.

Table 4.4 : Truth table

Inputs			Outputs							
A_2	A_1	A_0	y_7	y_6	y_5	y_4	y_3	y_2	y_1	y_0
0	0	0	0	0	0	0	0	0	0	1
0	0	1	0	0	0	0	0	0	1	0
0	1	0	0	0	1	0	0	1	0	0
0	1	1	0	0	0	0	1	0	0	0
1	0	0	0	0	0	1	0	0	0	0
1	0	1	0	0	1	0	0	0	0	0
1	1	0	0	1	0	0	0	0	0	0
1	1	1	1	0	0	0	0	0	0	0

```
Library ieee;
Use ieee.std_logic_1164.all;
     ieee.numeric_std.all;
Use ieee.std_logic_arith.all;
Use ieee.std_logic_unsigned.all;
entity Decoder 3_8 is
    port (A : in integer range 0 to 7;
          y : out std_logic_vector (7 downto 0));
end Decoder 3_8;
architecture DECODE of Decoder 3_8 is
    begin
        process (A)
            begin
                Case A is
                    When 0 => y <= "00000001";
```


Fig. 4.8

```
            When 1 => y <= "00000010";
            When 2 => y <= "00000100";
            When 3 => y <= "00001000";
            When 4 => y <= "00010000";
            When 5 => y <= "00100000";
            When 6 => y <= "01000000";
            When 7 => y <= "10000000";
        end case;
    end process;
end DECODE;
```

4.4 MOORE FINITE STATE MACHINE

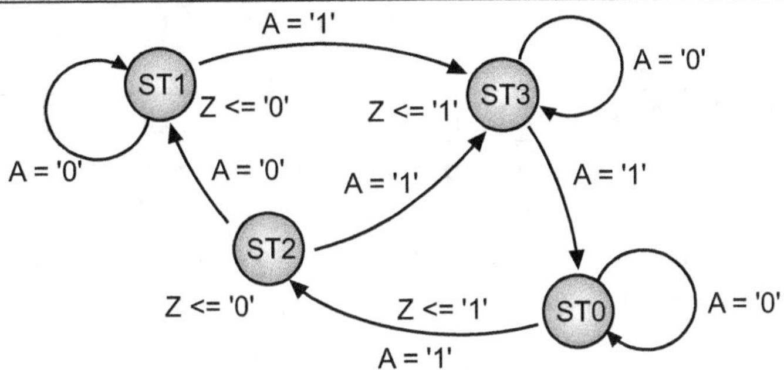

Fig. 4.9

- The output of a Moore finite-state-machine (FSM) depends only on the state and not on its inputs.
- This type of behavior can be modeled using a single process with a case statement that switches on the state value.
- State transition diagram for Moore FSM is shown below:

```
Library ieee;
Use ieee.std_logic_1164.all;
    entity Moore_FSM is
        port (A, CLK: in bit;
            Z : out std_logic);
    end Moore_FSM;
architecture FSM_exam of Moore_FSM is
type state_type is (ST0, ST1, ST2, ST3);
Signal Moore_State : State_type;
    begin
        process (CLK)
        begin
            if CLK = '0' then
            Case Moore_state is
            When ST0 => Z <= '1';
            if A = '1' then
                Moore_state <= ST2;
            end if;
            When ST1 => Z <= '0';
            if A = '1' then
                Moore_state <= ST3;
            end if;
            When ST2 => Z <= '0';
```

```
        if A = '0' then
                Moore_state <= ST1;
        else
                Moore_state <= ST3;
        end if;
        When ST3 => Z <= '1';
        if A = '1' then
                Moore_state <= ST0;
        end if;
    end case;
  end if;
  end process;
end FSM_exam;
```

4.5 MEALY FINITE STATE MACHINE

- In a Mealy finite state machine, the outputs not only depend on the state of the machine but also on its inputs.
- This type of finite state machine can also be modeled in a style similar to that of the Moore machine with case statement using single process.
- To show another style.
  In this case we use two processes,
- One process that models the synchronous aspect of FSM, and one that models the combinational part of the FSM.

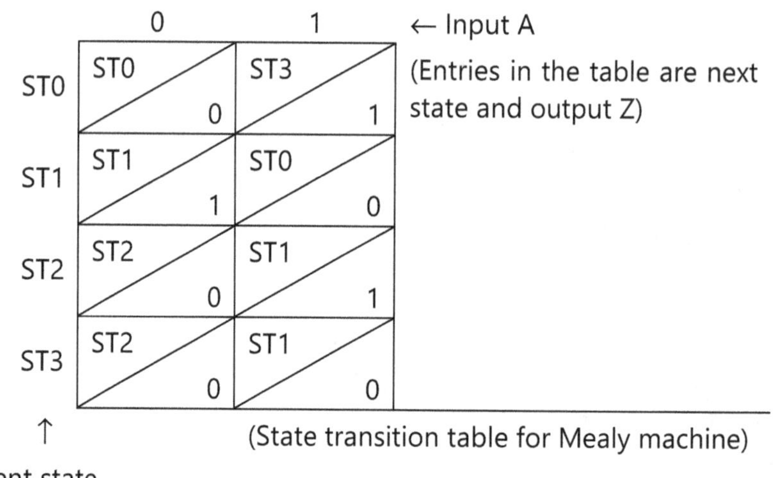

Fig. 4.10

```
library ieee;
Use ieee.std_logic_1164.all;
    entity Mealy_FSM is
        port (A, CLK : in bit;
            Z : out std_logic);
    end Mealy_FSM;
    architecture FSM_exam of Mealy_FSM is
        type Mealy_type is (ST0, ST1, ST2, ST3);
        Signal P_State, N_State : Mealy_type;
    begin
        Seq_part : Process (CLK)
```

```vhdl
        begin                                    -- Synchronous-section
            if CLK = '0' then
                P_state <= N_State;
            end if;
        end process Seq_part;
    Comb_part : process (P_State, A)
        begin
            case P_State is
                when ST0 =>
                    if A = '1' then
                        Z <= '1';
                        N_State <= ST3;
                    else
                        Z <= '0';
                    end if;
                when ST1 =>
                    if A = '1' then
                        Z <= '0';
                        N_state <= ST0;
                    else
                        Z <= '1';
                    end if;
                when ST2 =>
                    if A = '0' then
                      Z <= '0'
                    else
                       Z <= '1';
                        N_state <= ST1;
                end if;
                when ST3 =>
                    Z <= '0';
                if A = '0' then
                    N_State <- ST2;
                    else
                    N_State <= ST1;
                    end if;
                end case;
            end process comb_part;
        end FSM_exam;
```

- In this type of FSM, it is important to put the input signals in the sensitivity list for the combinational part process, since the outputs may directly depend on the inputs independent of the CLK.
- Such condition does not occur in a Moore FSM since output depends only on states.

4.5 MULTIPLEXERS

4.5.1 Introduction

- A multiplexer allows digital signals from several sources to be routed onto a single bus or line. A 'select' input to the microprocessor allows the source of the signal to be chosen.

4.5.2 2 : 1 Multiplexer

1. Block Diagram :

- The 2 : 1 multiplexer is shown in Fig. 4.11. A logic '1' on the SEL line will connect the 4-bit input bus A to the 4-bit output bus X. A logic '0' on the SEL line will connect input bus B to output bus X.

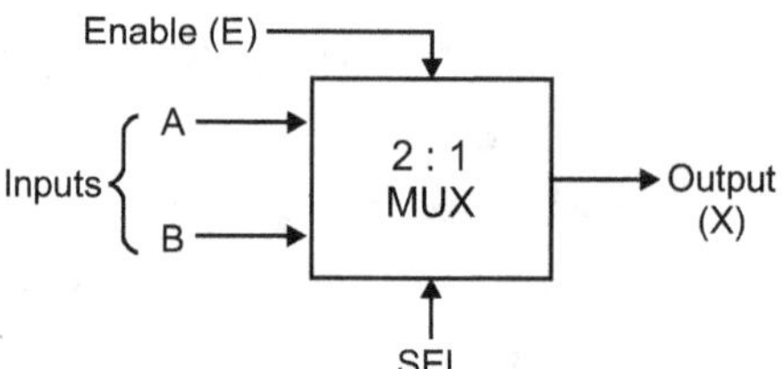

Fig. 4.11 : 4-bit 2 : 1 multiplexer

2. VHDL Code :

- The VHDL code for implementing the 4-bit 2 : 1 multiplexer is shown below :

```
Library IEEE;
use IEEE.STD_LOGIC_1164.ALL;
entity mux_2 to 1_top is
     Port(SEL :  in STD_LOGIC;
             A :  in STD_LOGIC_VECTOR (3 downto 0);
             B :  in STD_LOGIC_VECTOR (3 downto 0);
             X :  out STD_LOGIC_VECTOR (3 downto 0);
end mux_2 to 1_top;
architecture Behavioral of mux_2 to 1_top is
begin
     X < = A when (SEL = '1'), else B;
end behavioral;
```

3. Description :

- The VHDL *when* and *else* keywords are used to implement the multiplexer.
- The *when-else* construct is a conditional signal assignment construct that assigns the signal on the left of *when* A to the output signal X, if the condition to the right of *when* is true (SEL '1'- if SEL is equal to logic '1').
- If the condition is false, then the signal to the right of the else B will be assignment to the output signal instead (this will occur if the signal on SEL is a logic '0').
- The result of this signal assignment is that a logic '1' on SEL will connect the 4-bit input bus A to the 4-bit output bus X. A logic '0' on SEL will connect the 4-bit input bus B to the output bus X.

4. Source Code :

- The source file for the 4-bit 2 : 1 multiplexer can be downloaded. The UCF and JED files are configured for use on the home made CPLD board.
- The VHDL, UCF and JFD files : mux_2 to 1_4 bit_zip (6.1 kB).

4.5.3 4 : 1 Multiplexer

1. Block Schematic :

- A 1-bit wide 4 : 1 multiplexer that multiplies single (1-bit) signals is shown in Fig. 4.12. The two SEL pins determine which of the four inputs will be connected to the output.

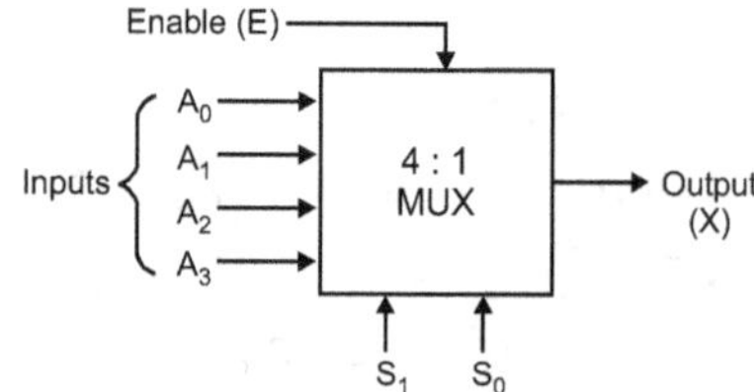

Fig. 4.12 : 1 bit 4:1 MUX

2. VHDL Code :

- The VHDL code that implements the 1-bit 4 : 1 multiplexer is shown below :

```
Library IEEE;
use IEEE.STD_LOGIC_1164.ALL;
entity mux_4 to 1_top is
```

```
        Port(SEL : in STD_LOGIC_VECTOR (4 downto 0);
              A : in STD_LOGIC_VECTOR (3 downto 0);
              X : out STD_LOGIC;
        end mux_4 to 1_top;
        architecture Behavioral of mux_4 to 1_top is
        begin
        with SEL select
              X = A(0) when '00';
                  A(1) when '01';
                  A(2) when '10';
                  A(3) when '11';
                  '0' when others;
        end behavioral;
```

3. Logic Circuit :

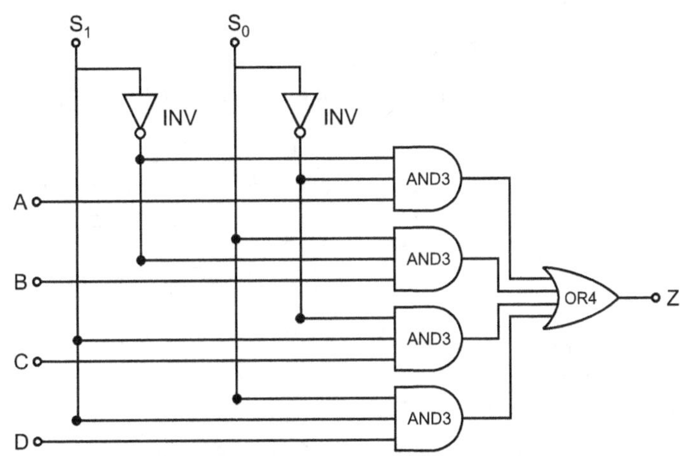

Fig. 4.13 : Logic Circuit of 4 : 1 MUX

- The logic circuit of a 4 : 1 binary multiplexer using logic gates AND and OR is shown in Fig. 4.13.

4. Truth Table :

Table 4.5

Inputs		Output
S_1	S_0	Z
0	0	A
0	1	B
1	0	C
1	1	D

- The truth table of a 4 : 1 multiplexer is shown in Table 4.5.

5. Description :

- In this code, *with, select* and *when* VHDL key words are used.
- The line *with SEL select* sets up SEL as the signal that is evaluated by the *when* statements that follow when the logic levels on SEL match one of the refer to the right of one of the *when* statements, the signal to the left of that *when* statement will be assigned to the output signal X.
- The line containing *'others'* is required by VHDL to take case of any logic combination that is not taken cases of by the preceding statements. This allows for any states besides logic 0 and logic 1 levels, such as high impedance signal-Z.
- Note that single signals are assigned logic values by using single quotes, e.g. '0'. A group of signals (vectors) are assigned values between double quotes, e.g. "1".
- VHDL multiplexers can be implemented by using when-else construct code.

4.6 ENCODERS

4.6.1 Definition

- **An encoder is a combination of logic circuit that takes in multiple inputs, encodes them and outputs an encoded version with fewer bits.**
- In a simple encoder, only one of the input lines in active at any moment. In this post, we will write the VHDL code for a 4 : 2 encoder using its logic equations and its truth table.

4.6.2 Binary Encoders

- Binary encoders has 2n inputs and n-bit output lines. It can be 4 : 2, 8 : 3 and 16 : 4 line configurations.

4.6.3 4 : 2 Encoder

1.　Block Schematic :

- The 4 : 2 binary encoder is shown in Fig. 4.14. It has four input lines and two output lines.

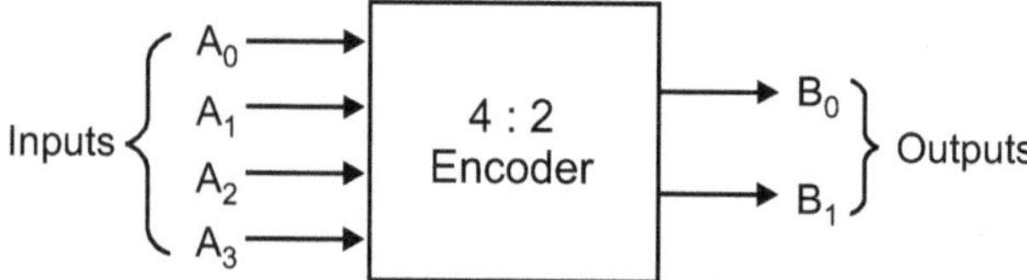

Fig. 4.14 : 4 : 2 encoder

2.　Logic Circuit :

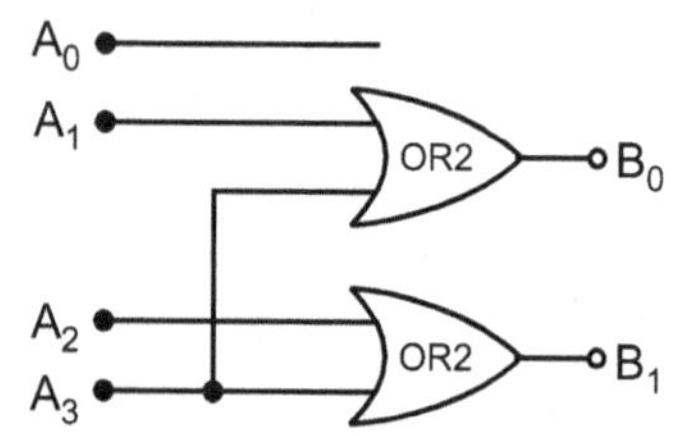

Fig. 4.15 : Logic Circuit of 4 : 2 encoder

Table 4.6

Inputs				Outputs	
A_3	A_2	A_1	A_0	B_1	B_0
0	0	0	1	0	0
0	0	1	0	0	1
0	1	0	0	1	0
1	0	0	0	1	1

- The logic circuit of a 4 : 2 binary encoder is shown in Fig. 4.15. It uses logic gates.

3.　Truth Table :

- The truth table of a 4 : 2 binary logic is shown in Table 4.6.

4.　VHDL Code :

- VHDL code for 4 to 2 binary encoder can be designed both in structured and behavioral modeling.
- VHDL code for 4 to 2 encodes can be done in different methods like using case statement using of else statement, using logic gates etc. Here we provide example code for all 3 methods for better understanding of the VHDL language.

(a) Using 'case' statement :

```
Library IEEE;
use
IEEE.STD_LOGIC_1164.all;
entity encodes is
    port(
        a : in STD_LOGIC_VECTOR(3 downto 0);
        b : out STD_LOGIC_VECTOR(3 downto 0);
        );
end encoder;
architecture bhv of encoder is
begin
process (a)
begin
    case a is
    when "1000" => b <= "00";
```

```vhdl
            when "0100" => b <= "01";
            when "0010" => b <= "10";
            when "0001" => b <= "11";
            when others => b <= "11";
        end case;
        end process;
    end bhv;
```

(b) Using 'if' statement :

```vhdl
    Library IEEE;
    use
    IEEE.STD_LOGIC_1164.all;
    entity encoder 1 is
        port(
                a : in STD_LOGIC_VECTOR(3 downto 0);
                b : out STD_LOGIC_VECTOR(1 downto 0);
                );
        end encoder 1;
        architecture bhv of encoder is
        begin
        process (a)
                if (a = "1000") then b <= "00";
            elseif (a = "0100") then b <= "01";
            elseif (a = "0001") then b <= "11";
            elseif b <= "zz";
        end if;
    end process;
    end bhv;
```

(c) Using 'Logic' gates :

```vhdl
    Library IEEE;
    use
    IEEE.STD_LOGIC_1164.all;
    entity encoder 2 is
        port(
                a : in STD_LOGIC_VECTOR(3 downto 0);
                b : out STD_LOGIC_VECTOR(1 downto 0);
                );
    end encoder 2;
    architecture bhv of encoder 2 is
    begin
            b(0) < = a (1) or a(2);
            b(1) < = a (1) or a(3);
    end bhv;
```

4.7 DECODERS

4.7.1 Definition

- **A decoder is a combinational logic circuit that does the opposite job of an encoder.**

4.7.2 Binary Decoder

- Binary decoder has n-bit input times and 2 power n output lines. It can be 2 to 4, 3 to 8, and 4 to 16 line configurations. Binary decoder can be easily constructed by using basic logic gates. VHDL code of 2 to 4 decoder can be easily implemented with structural and behavioral modeling.

- A decoder that has two inputs, an enable pin, and four outputs is implemented in a CPLD using VHDL.
- This 2 to 4 decoder will switch on one of the four active low outputs, depending on the binary value of the two inputs and if the enable input is high.

4.7.3 2 : 4 Decoder

1. Block Schematic :

- The block schematic of a 2 : 4 binary decoder is shown in Fig. 4.16.

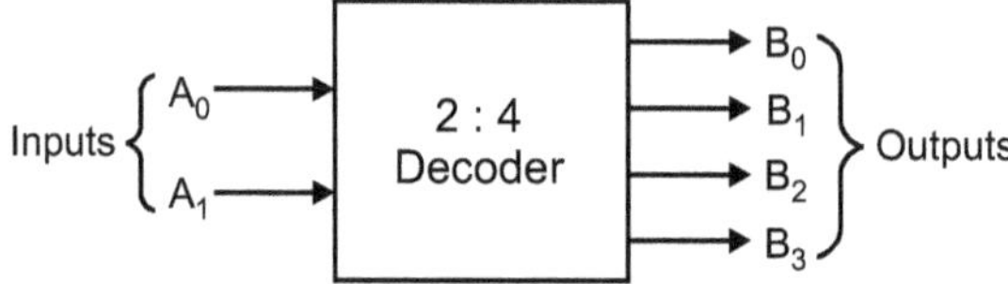

Fig. 4.16 : 2 : 4 decoder

2. Logic Circuit :

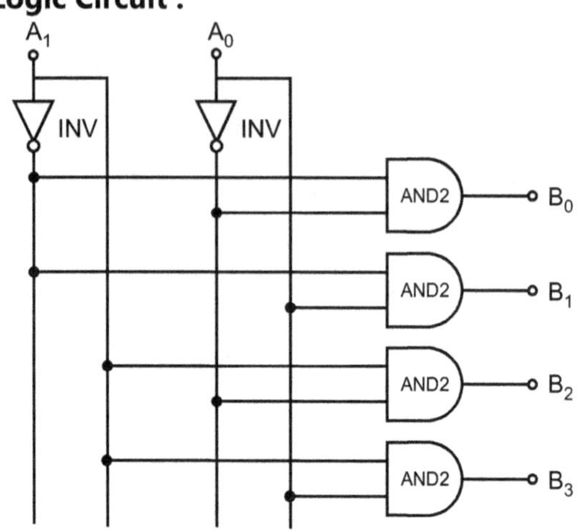

Fig. 4.17 : 2 : 4 decoder using logic gates

Table 4.7

Inputs		Outputs			
A_1	A_0	B_3	B_2	B_1	B_0
0	0	0	0	0	1
0	1	0	0	1	0
1	0	0	1	0	0
1	1	1	0	0	0

- The logic circuit of a 2 : 4 decoder using logic gates is shown in Fig. 4.17.

3. Truth Table :

- The truth table for a 2 : 4 decoder is shown in Table 4.7.

4. VHDL Code :

- Here we provide example code for all three methods for better understanding of the VHDL language.

  (a) Using 'case' statement :

```vhdl
Library IEEE;
use
IEEE.STD_LOGIC_1164.all;
entity decoder is
    port(
            a : in STD_LOGIC_VECTOR(1 downto 0);
            b : out STD_LOGIC_VECTOR(3 downto 0);
        );
end encoder;
architecture bhv of decoder is
begin
process (a)
begin
        case a is
        when "00" => b <= "0001";
        when "01" => b <= "0010";
```

```vhdl
            when "10" => b <= "0100";
            when "11" => b <= "1000";
            when others => b <= "11";
end case;
end process;
end bhv;
```

(b) Using 'if else' statement :

```vhdl
Library IEEE;
use
IEEE.STD_LOGIC_1164.all;
entity decoder 1 is
    port(
            a : in STD_LOGIC_VECTOR(1 downto 0);
            b : out STD_LOGIC_VECTOR(3 downto 0);
        );
    end decoder 1;
    architecture bhv of decoder 1 is
    begin
    process (a)
    begin
            if (a = "00") then b <= "0001";
            elseif (a = "01") then b <= "0010";
            elseif (a = "10") then b <= "0100";
            elseif (a = "11") then b <= "1000";
    end if;
    end process;
end bhv;
```

(c) Using 'Logic' gates :

```vhdl
Library IEEE;
use
IEEE.STD_LOGIC_1164.all;
entity decoder 2 is
    port(
            a : in STD_LOGIC_VECTOR(1 downto 0);
            b : out STD_LOGIC_VECTOR(3 downto 0);
        );
    end decoder 2;
    architecture bhv of decoder 2 is
    begin
            b(0) < = not a (0) and a (1);
            b(1) < = not a (0) and a (1);
            b(2) < = a (0) and not a(1);
            b(3) < = a (0) and not a(1);
end bhv;
```

4.8 IMPLEMENTATION OF BOOLEAN FUNCTIONS

4.8.1 Introduction

- A logic variable can take one of the two values, typically called *true* and *false* or *high* or *low*.
- There are two types of logic circuits depending on Boolean functions, namely *combinational logic* and *sequential logic*.
- In addition to the standard logic functions (AND, OR, NOT, XOR, NAND, etc.), combinational logic functions that are widely used include multiplexer, encoder and decoder.
- In addition to the standard logic functions (flip-flop), some sequential logic functions that are widely used include microprocessor, computer, shift register, counter etc.

4.8.2 Shift Registers

1. **Definition :**

- *A shift register is a circuit of several flip flops where the output of each flip-flop is connected to the input of the adjacent flip-flop.*
- The sequential circuit will change the state on every rising edge of the clock signal.

2. **Types :**

- The shift register can be categorised as :

  1. Serial-in series-out (SISO),
  2. Serial-in parallel-out (SIPO),
  3. Parallel-in parallel-out (PIPO)
  4. Parallel-in serial-out (PISO) shift registers.

4.8.3 VHDL Code for PIPO Shift Register

```vhdl
library ieee;
use ieee.std_logic_1164.all;
entity pipo is
port(
    clk : in std_logic;
    D: in std_logic_vector(3 downto 0);
    Q: out std_logic_vector(3 downto 0)
    );
end pipo;
architecture arch of pipo is
begin
process (clk)
begin
    if (CLK'event and CLK='1')
    then
    Q < = D;
    end if;
    end process;
end arch;
```

4.8.4 VHDL Code for SIPO Shift Register

```vhdl
library ieee;
use ieee.std_logic_1164.all;
entity sipo is
port(
    clk, clear : in std_logic;
    Input_Data: in std_logic;
    Q: out std_logic_vector(3 downto 0) );
end sipo;
architecture arch of sipo is
```

```vhdl
begin
process (clk)
begin
    if clear = '1' then Q <= "0000";
    elsif (CLK'event and CLK='1') then
    Q(3 downto 1) < = Q(2 downto 0);
    Q(0) < = Input_Data;
    end if;
    end process;
end arch;
```

4.9 COUNTERS

4.9.1 Definition

- *A counter is a device which stores the number of times a particular event or process has occurred, often in relationship to a clock signal.*

- There are two types of counters, namely up counter and down counter and also up-down counter.

4.9.2 VHDL Code for Up Counter

```vhdl
library IEEE;
use IEEE.STD_LOGIC_1164.ALL;
use IEEE.STD_LOGIC_UNSIGNED.ALL;
entity UP_COUNTER is
    Port    ( clk: in std_logic;
            reset: in std_logic;
            counter: out std_logic_vector(3 downto 0)
            );
end UP_COUNTER;
architecture Behavioral of UP_COUNTER is
signal counter_up: std_logic_vector(3 downto 0);
begin
    if(rising_edge(clk)) then
    if(reset='1') then
        counter_up <= x"0";
    else
        counter_up <= counter_up + x"1";
    end if;
    end process;
        counter <= counter_up;
end Behavioral;
```

4.9.3 VHDL Code for Down Counter

```vhdl
library IEEE;
use IEEE.STD_LOGIC_1164.ALL;
use IEEE.STD_LOGIC_UNSIGNED.ALL;
entity DOWN_COUNTER is
    Port    ( clk: in std_logic;
            reset: in std_logic;
            counter: out std_logic_vector(3 downto 0)
             );
```

```vhdl
end DOWN_COUNTER;
architecture Behavioral of DOWN_COUNTER is
signal counter_down: std_logic_vector(3 downto 0);
begin
process(clk,reset)
begin
    if(rising_edge(clk)) then
        counter_down <= x"F";
     else
        counter_down <= counter_down - x"1";
    end if;
end if;
end process;
    counter << counter_down;
end Behavioral;
```

4.9.4 VHDL Code for Up-Down Counter

```vhdl
library IEEE;
use IEEE.STD_LOGIC_1164.ALL;
use IEEE.STD_LOGIC_UNSIGNED.ALL;
entity UPDOWN_COUNTER is
    Port ( clk: in std_logic;
            reset: in std_logic;
            up_down: in std_logic;
            counter: out std_logic_vector(3 downto 0)
            );
end UPDOWN_COUNTER;
architecture Behavioral of UPDOWN_COUNTER is
signal counter_updown: std_logic_vector(3 downto 0);
begin
    process(clk,reset)
begin
    if(rising_edge(clk)) then
        if(reset='1') then
            counter_updown <= x"0";
    elsif(up_down='1') then
        counter_updown <= counter_updown - x"1";
    else
        counter_updown <= counter_updown + x"1"; -- count up
    end if;
end if;
end process;
    counter <= counter_updown;
end Behavioral;
```

4.10 TEST BENCH IN VHDL

4.10.1 Introduction

- An engineer's job does not stop after having created 'solution' to a specific problem, but he/she must also be able to demonstrate, to various degrees of centridude, that the solution is correct. This statement is equally valid in both software and hardware domain, an errors can generally not be fixed once a product has been shipped to customers!

- When describing digital circuits in VHDL, one generally tests the correctness of their implementation with a VHDL test bench, which is generally *non-synthesizable* VHDL file, which is iteratively applie a sequence of controlled inputs to a circuit and compares its concrete output against the operated output.

- If a mismatch is detected, an error is displayed in the VHDL simulators log which can then be consulted to help directed designer search for the problem in the circuit's RTL description.

- This tutorial introduces readers to the craft of writing simple VHDL testbenches. We start with an empty VHDL test bench file and iteratively explains our thought process and it affects the way we construct the testbench. We start with a simple testbench for a combinational circuit, then move on towards a more complicated testbench for a sequential circuit.

- Test bench is an important part of VHDL design to check the functionality of design through the simulation waveform. It provides stimulus for design under test (DUT) or unit under test (UUT) to check the output result.

- A test bench is HDL code that allows you to provide a documented, repeatable set of stimulii that is portable across different simulators. It consists of entity without any I/O ports, design is instantiated as compact, clock input, and various stimulus inputs.

4.10.2 Definition

- **The design which allows to verity functionality of design at each step in VHDL synthesis-based methodology is known as test bench.**

- Test bench is a piece of VHDL code, whose purpose is to verify the functional correctness of VHDL model.

- A **test bench** or **testing workbench** is an environment used to verify the correctness or soundness of a design or model.

- The test bench is a specification in VHDL that plays the role of a complete set of stimuli that is portable across different simulators.

4.10.3 Test bench Syntax

ENTITY tb_name IS END tb_name;
ARCHITECTURE tb_arch OF tb_name
IS
Component Declaration **for** the
Unit Under Test (UUT)
Input/Output Signal Declaration
Clock period definitions
BEGIN
Instantiate the Unit Under Test (UUT)
Clock **process** definitions
Stimulus **process**
END tb_arch;
VHDL Testbench syntax consists of Header File declaration containing library
LIBRARY ieee;
USE ieee.std_logic_1164.ALL;
entity without ports declaration
ENTITY tb_up_down IS
END tb_up_down;
architecture with component declaration for unit under test

4.10.4 Testbench Architecture

- *Software application process of defining a structured solution that meets all of the technical and operational requirements, while optimizing common quality attributes such as performance, security and manageability.*

- It involves a series of decisions based on a wide range of factors and each of these decisions can have considerable impact on the quality, performance, maintainability, and overall success of the application. It consists of four process steps.

- Fig. 4.18 shows the architecture of a VHDL testbench as under :
  1. Prerequisite
  2. Structure
  3. Validity
  4. Delivery

Step 1: Prerequisite

- The first step is understanding the prerequisite, which set the requirements for the architecture. There are four key prerequisites for the test bench architecture process as under :
  1. Design under test (DUT)
  2. Verification plan
  3. Application world abstraction
  4. Reuse strategy

- It might seen tempting to directly jump into the test bench architecture with partial understanding of the prerequisites or even skipping some.

- Especially it is tempting to skip writing the verification plan. It takes time and effort to come up with a good verification plan containing test plan, coverage plan and check plan but do not skip it. This is one of the very important prerequisites, as the primary objective of test bench architecture is to meet the requirements of the verification plan.

- Verification plan also serves as a means to validate the architecture. Bottom line, **test bench is the means to execute verification plan**. So a test bench architecture without a verification plan is strictly no.

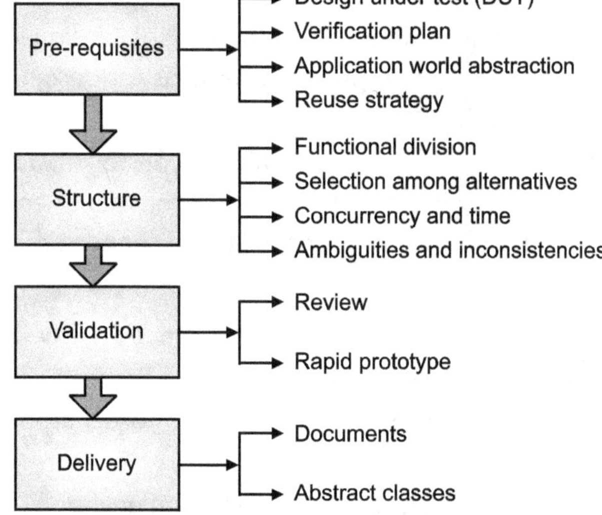

Fig. 4.18 : Architecture of test-bench 4-step process

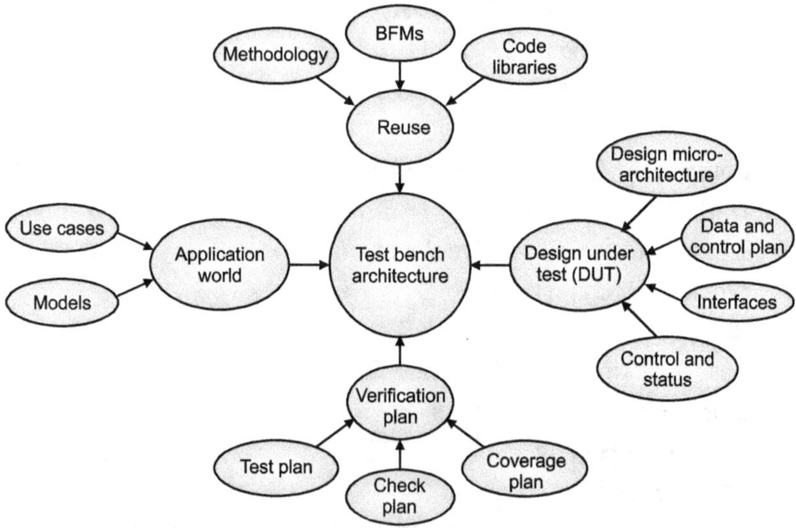

Fig. 4.19 : Testbench architecture 4-key driving factors

Step 2: Structure

- The second step is dividing the functionality to various components meeting the requirements and mapping to the classes.

- After all preparations, when a great architecture emerges out, it almost seems like a magic. It just beautifully handles all the cases, even the cases that were not anticipated. It is a pure joy to watch it in action.

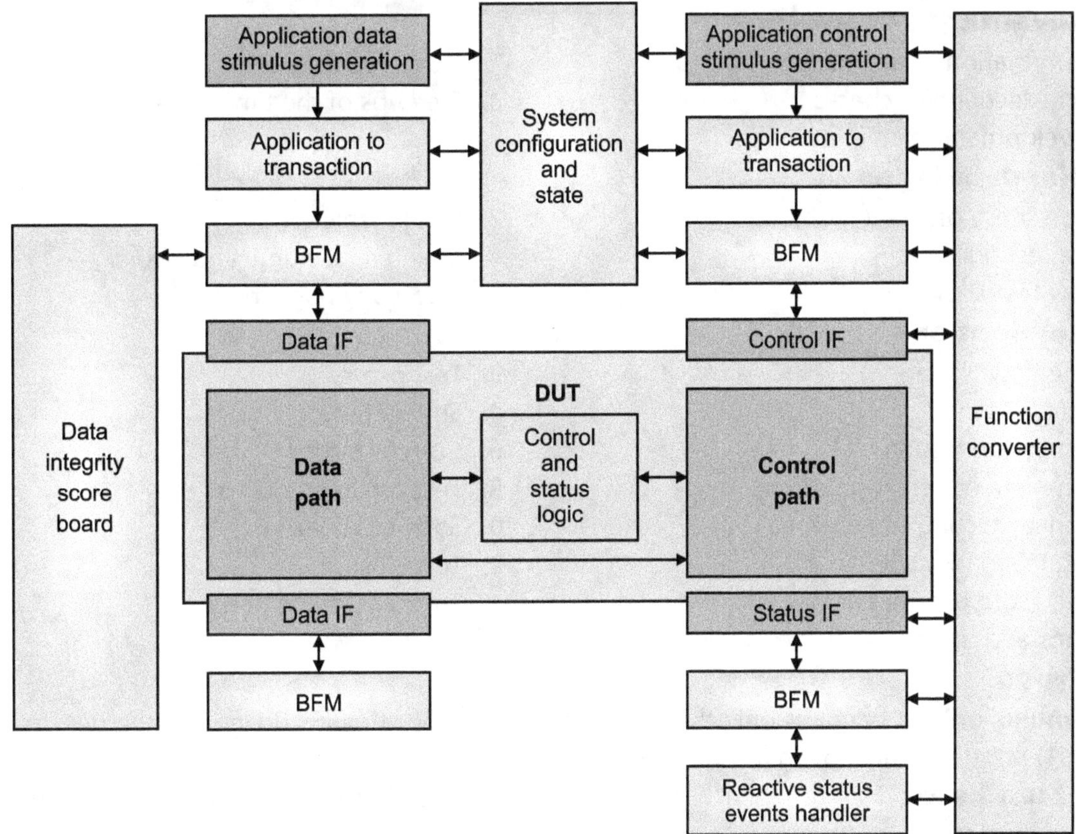

Fig. 4.20 : Structural view of testbench architecture

- Fig. 4.20 shows the structural view of testbench architecture.
- Following are the steps to completely build the test bench architecture.
  1. Functional division
  2. Selection among alternatives
  3. Handling concurrency and time
  4. Handling ambiguity and inconsistency

Step 3: Validation

- The third step is validation, to check if the architecture has any inconsistencies.
- Test bench architecture validation is a process to check if it meets the verification requirements. Additionally, check for inconsistencies and complexity. In some cases, complexity may be necessary, but validation should ensure that complexity is manageable.
- This can be done with the help of review and rapid prototyping.

Step 4: Delivery

- Fourth and the final step are architecture delivery to execution by packaging it in documents and abstract classes.
- Ideal test bench architecture hand off from architect to execution team should be through a set of abstract classes. **Abstract classes with the tasks, functions and key threads are identified.** Key user exposed APIs defined. Comments in header defining the functionality of class and functionality for tasks and functions to be written for guidance should be provided. Bonus points to architect for stitching these abstract classes and get it compiling before passing to execution team.
- This is one of the best ways to pass intent without diluting. This allows restricting any poor implementation by the execution team to a class, function or task rather than letting it spread throughout like infectious disease all over the test bench code. This enables recovery from the damages of poor coding or lack of expertise of the execution team. It is more work, but allows to preserve the cleanliness and intent.

- There may be some practical limitations due to bandwidth availability of the architect. In this case the architect should first get these abstract classes implemented by the execution team and review them before the full blown execution kicks off.

4.10.5 Advantages

1. Confirmation of the known.
2. Quick feedback.
3. Fast execution of checks.
4. Free ups of the time of testers.
5. Development team contribution.

4.10.6 Disadvantages

1. False sense of quality.
2. Not reliable.
3. Not testing.
4. Maintenance time and effort.
5. Slow feedback.
6. Not many bugs found.

4.10.7 Applications

1. Data flow testing,
2. Test process
3. Sanity testing
4. Smoke testing
5. Regression testing
6. Current testing
7. Complete selection guide
8. Test care's pretiration
9. Monkey testing
10. Software review
11. Rotate string guide
12. Test closure
13. Active and passive testing
14. Entry and exit criteria
15. Stubs and drivers

4.10.8 Types

1. **Stimulus only:** It contains only the driver containing the stimulus driver and the design under test (DUT) but does not contain any result verification.
2. **Full test bench:** This test bench contains the stimulus driver, the correct results and results for comparison.
3. **Simulator specific test bench:** The name suggests that the test bench is written in a simulator specific format.
4. **Hybrid test bench:** This is a blend of techniques from more than one test bench style.
5. **Fast test bench:** It is to optimize the speed of a test bench. This is written in a way so as to provide the best speed from a simulation.

4.10.9 Examples of Applications

- A testbench is used to verify the functionality of the design.

Test_Bench in VHDL :

- The language provides a large number of ways to write a test bench.
- A typical testbench format is,

```
entity TEST_BENCH is
end;
architecture TB_BEHAVIOR of TEST_BENCH is
    Component ENTITY-UNDER_TEST
        port (list-of-ports-their types-and-modes);
    end component;
        Local-signal declarations;
    begin
        Generate-waveforms-using-behavioral - construct;
        Apply-to-entity-under-test;
        EUT : ENTITY_UNDER_TEST port map (port associations);
        Monitor-values-and-compare;
    end TB_BEHAVIOR;
```

[Stimulus is automatically applied to the entity under test by instantiating the entity in the testbench model and then specifying the appropriate interface signals.]

Example : Test-bench of AND Gate :

Note :

1. A testbench is a program that defines a set of input signals to verify the operation of a circuit. e.g. AND Gate.

2. The testbench takes no inputs and returns no outputs. As such as ENTITY declaration is empty.

3. The circuit under verification, here the AND Gate, is imported into the testbench **Architecture** as a component.

```
library ieee;
use ieee.std_logic_1164.all; - - import std_logic from the IEEE library
entity And Gate_TB is
end And Gate_TB; - - Entity declaration : no inputs, no outputs.
architecture tb of And Gate_TB is - - Describe how to test the AND Gate.
Component And Gate is
    port (A, B : in std_logic; - - pass And Gate entity to the testbench as component.
          Y : out std_logic);
end component;
    Signal in A, in B, out Y : std_logic;
begin
    mapping : And Gate port map (in A, in B, out y);
                                        - - map the testbench signals to the And Gate
Process
    Variable errcnt : integer ;=0; - - Variable to track errors.
begin - - Test-1
    in A <= '0';
    in B <= '0';
    wait for 15 ns;
    assert (out Y = '0') report "Error 1"
    if (out Y/= '0') then
        errcnt := errcnt + 1;
    end if;
        in A <= '0'; - - TEST-2
        in B <= '1';
        wait for 15 ns;
        assert (out Y = '0') report 'Error 2"
        if (out Y/= '0') then
        errcnt := errcnt + 1;
    end if;
        in A <= '1';
        in B <= '1';
        wait for 15 ns;
        assert (out Y = '1') report "Error-3"
        if (out Y /= '1') then
        errcnt := errcnt + 1;
    end if;
```

.... Summary ....

```
        if (errcnt = 0) then
            assert false report "Good" severity note;
        else
            assert true report "Error' severity error;
        end if;
    end process;
end tb;
```

....

```
configuration cfg_tb of And Gate_tb is
        for tb
    end for;
end cfg_tb;
```

.... END
.... END.

Practice Questions

1.   Explain 'ENTITY' and 'ARCHITECTURE' with syntax.
2.   Explain Data Flow type modeling with an example.
3.   List the different types of sequential statements and write the syntax of any one statement.
4.   Describe following statements with syntax: (i) Process statement (ii) Generate statement
5.   Explain the data types used in VHDL.
6.   Describe the modeling of conditional operators by taking one suitable example.
7.   Explain the following statements: (i) Next statement (ii) Wait statement
8.   What do you mean by testbench in VHDL design ?
9.   What are the applications of testbenches in VHDL ?
10. What is a combinational logic circuit ?
11. List some of the combinational logic circuits.
12. Design a 4 : 1 multiplexer using VHDL code.
13. What is a multiplexer ?
14. What is an encoder ?
15. Design a 4 : 2 encoder using VHDL code.
16. What is a decoder ?
17. Design a 2 : 4 binary decreases using VHDL code.
18. What is a code converter ?
19. Design a BCD to Excess-3 counters using VHDL code.
20. What is an excess-3 code ?
21. What is gray code ?
22. Design a Gray to Binary code convertor using VHDL code ?
23. What is a binary code ?
24. What is a combination logic circuit ?
25. What is a sequential logic circuit ?
26. List some of the sequential logic circuits.
27. What is a shaft register ?
28. Design a PIPO shift register using VHDL code.
29. Design a SIPO shift register using VHDL code.
30. What is a counter ? What are its types ?
31. Design up-counter using VHDL code.
32. Design down counter using VHDL code.
33. Design up-down counter using VHDL code.

Chapter 5

HDL SIMULATION AND SYNTHESIS

Syllabus

5.1 Event scheduling, Sensitivity list, Zero modeling, Simulation cycle, Comparison of software and hardware description language, Delta delay.

5.2 HDL design flow for synthesis.

5.3 Efficient coding styles, Optimizing arithmetic expression, Sharing of complex operator.

Learning Objectives

(5a) Describe VHDL simulation for the given application.

(5b) Draw HDL design flow of synthesis for the given application.

(5c) Describe use of efficient coding styles, Optimizing expression, Sharing of complex operator.

5.1 SIMULATION

5.1.1 Introduction

- Now-a-days, the economy requires a fast and flexible reaction to the market. Customer demands become more and more dynamic and unpredictable. The only tool that can analyze and improve this complex and dynamic system is simulation.
- Simulation is the imitation of the operation of a real-world process or system over time. The act of simulating something first requires that a model be developed; this model represents the key characteristics or behaviors of the selected physical or abstract system or process.
- The model represents the system itself, whereas the simulation represents the operation of the system over time.
- Key issues in simulation include the acquisition of valid source information about the relevant selection key characteristics and behaviors, the use of simplifying approximations and assumptions within the simulation, and fidelity and validity of the simulation outcomes.
- Procedures and protocols for model verification and validation are an ongoing field of academic study, refinement, research and development in simulations technology or practice, particularly in the field of computer simulation.

5.1.2 Definition

- *A simulation is an approximate imitation of the operation of a process or system that represents its operation over time.*
- Simulation is used in many contexts, such as simulation of technology for performances optimization, safety engineering, testing, training, education, and video games.
- Simulation is a model of a set of problems or events that can be used to teach some one how to do something, or the process of making such a model.
- **A simulation is any research or development project, where researches or developers create a model of some authentic phenomenon.**

5.1.3 Need

- Due to following reasons there is a need of simulation VHDL:
  1. Aids in discovering and removing waste
  2. Optimizing system and increase revenue
  3. Helps management
  4. Visualize environment
  5. Proposed changes
  6. Right tool (simulator software)

5.1.4 Advantages

1. Test the behavior
2. Analyze the system
3. Give accurate results
4. Get faults in the system
5. Animation and modeling
6. Help students

5.1.5 Disadvantages

1. Expensive
2. Interpretation of simulation data
3. No optimal solution
4. Takes long time
5. Cannot simulate earthquake

5.1.6 Examples

1. Simulation of aircraft before making its first flight
2. Simulation of handling customers in the bank
3. Simulation of cars, airplanes, and other expensive automobiles
4. Testing of atomic bombs and its effects
5. Weather forecast
6. Simulation of traffic on roads
7. Ticket reservation on railway, airports
8. Simulation for education purposes
9. Simulation of building games
10. Exercise simulation

5.1.7 Areas of Application

1. Manufacturing
2. Semiconductor manufacturing
3. Construction engineering
4. Military application
5. Logistics, supply chains and distribution
6. Transportation modes and traffic
7. Business process
8. Healthcare
9. Automated material handling system (AMHS)
10. Test beds for functional testing of control system software
11. Risk analysis (Insurance, Portfolio, etc.)
12. Computer simulation (CPU, memory, etc.)
13. Networks (internet, backbone, LAN, switch, router, wireless, PSTN, call center etc.)
14. Bioinformatics
15. Classroom of the future
16. Communication satellite
17. Digital life cycle (CAD, CAM, CAE, etc.)
18. Disaster preparedness

5.1.8 Simulation Cycle

- First generation simulator used a technique. CAD developers called a one-list algorithm, which is relatively fast, but cannot handle parallel zero delay events such as exchanging A and B.

      A <= B; zero delay

      B <= A; zero delay

- This example would not exchange the value of A and B, but would give both A and B, the old value of B, using one-list algorithm.

- VHDL uses a two-list algorithm, which tracks the previous and new values of signals. In this method, expressions are first evaluated, then signals are assigned new values. In VHDL, the example code performs a data exchange between the two signals A and B at some point in simulation time.

- In operation, the old values of A and B are fetched and scheduled for assignment, for zero delay, after a subsequent WAIT statement is executed.

- The ordering of zero delay events is handled with a fictitious unit called delta time, which represents the execution of a simulation cycle without advancing simulation time.

- The key points of simulation and delta time are:

1. The simulator models zero-delay events using delta time.
2. Events scheduled at the same time are simulated in specific order during a delta time step.
3. Related logic is then re-simulated to propagate the effects for another delta time step.
4. Delta time steps continue until there is no activity for the same instant of simulated time.

5.2 EVENT SCHEDULING

5.2.1 Concept

- Event scheduling must take into account what impact particular dates of the event could have on the success of the event. When organizing a *scientific conference*, for example, organizers might take into account the knowledge in which periods classes are held at universities, since it is expected that many potential participants are university professors.

- The professors should also try to check that no other similar conferences are held at the same time, because overlapping would make a problem for those participants, who are interested in attending all conferences.

- When it is well known, who is expected to attend the event, organizers usually try to synchronize the time of the event with planned schedules of all participants. This is a difficult task, when there are many participants or when the participants are located at distant places. In such cases, the organizers should first define a set of suggested dates and address a query about suitable dates to potential participants. After response is obtained from all participants, the event time suitable for most of participants is selected. This procedure can be alleviated by internet tools.

5.2.2 Definition

- **Event scheduling is the activity of finding a suitable time for an event such as meeting, conference, trip, etc.** It is an important part of event planning that is usually carried out at its beginning stage.

- *The process of planning and coordinating the event is usually referred to as event planning.*

- **Event management is the application of project management to the creation and development of large-scale events such as festivals, conferences, ceremonies, weddings, formal parties, concerts, or conventions.**

5.2.3 Event Planning and Management

- Event planning can include budgeting, scheduling, site selection, acquiring necessary permits, coordinating transportation and parking, arranging the speakers or entertainers, arranging decor, event security, catering, coordinating with third party vendors, and emergency plans. Each event is different in its nature so process of planning and execution of each event differs on the basis of type of event.

5.2.4 Event Purpose and Concept

- Defining the purpose and concept of the event is an essential starting point of the event management process to ensure success. The event coordinating committee should brainstorm some aims and objectives of the event and ensure these are specific, measurable, achievable, realistic and have a time frame.

- The way in which an organization deals with events is known as event management. It includes the organization's objectives for managing events, assigned roles and responsibilities, critical success factors, standards, and event-handling procedures.

- The linkages between the various departments within the organization required handling events and the flow of this information between them is the focus of event management.

- Scheduling is used to allocate plan and machinery resources, plan human resources, plan production processes and purchase materials. In manufacturing, the purpose scheduling is to minimize, the production time and cost by tilling a production facility when to make, with which staff, and on which equipments.

5.2.5 Event Scheduling Algorithm

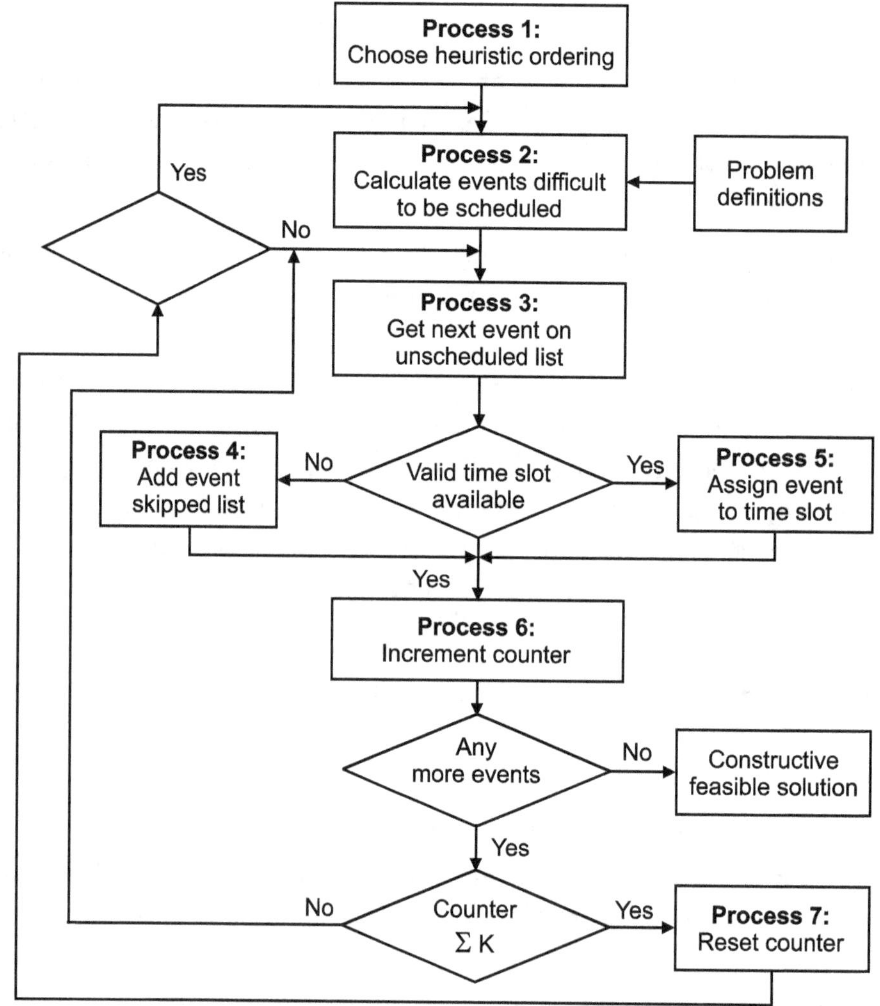

Fig. 5.1 : Modified Event Scheduling Algorithm

- A series of experiments were carried out to test the new modified algorithm of the event scheduling depicted in Fig. 5.1. Ultimately, the aim of experiments was to compare the solution quality when the different kinds of heuristics were employed to measure the difficulty of scheduling events to time slots.

5.2.6 Budgeting and Income

- Good financial management is fundamental to the delivery of successful events. Plan ahead be realistic, keep on top of the budget and implement control systems that 189 work for the event. In order to successfully plan the event an accurate and comprehensive budget should be created.

- As ticketing is an important means of crowd control, it should be considered a ticketing process that involves advanced ticket sales, tickets purchased at the event or both. Event management plan needs to address the following :
    1. Whether the tickets are pre-sold, sold at the gate or both;
    2. Information provided from the tickets about the tickets, about the event; and
    3. A description and/or a copy of the ticket.

5.2.7 Problem Statements/Constraints

(A) Hard constraints:
1. There should be only one event occurring at a time slot in a room.
2. Participants should participate at only one event at a given timeslot.
3. Volunteers should be unique for the event.
4. Coordinators should be unique for the event.

(B) Soft constraints:

1. Extension of the events may be permitted.
2. Cancellation of the events may occur when the stimulated number of participants is not there.
3. Postponement of the event may occur as the participant may be involved in other events.

- In the event scheduling, the events are scheduled so that there will not be any clash in the venue and time slots. This is done by group administrator. In login component, security measures are in built into each and every page, meaning that the private information will remain the same.

5.2.8 Benefits

1. No more missed appointments
2. Better time management
3. Greater organization
4. Monitor progress
5. Avoid scheduling
6. Motivation to accomplish goals
7. Streamlines from communication
8. Keep costs under controls
9. Preparation for unexpected
10. Serve time in communication
11. Efficiently manage multiple events

5.2.9 Advantages

1. Effective in a general purpose
2. Fair treatments for all processes
3. Load overhead on processor
4. Good response time for most processes
5. Flexible schedules

5.2.10 Disadvantages

1. Care must be taken in choosing quantum values.
2. Processing overhead is there in handling clock interrupt.
3. No process waits for more than (n-1) time units the next time quantum.

5.2.11 Events Queue

- Events are changes to the wire or registers. Statements can schedule event to occur at particular time or to be triggered by other events in current time slot or at later simulation time.
- The events queue is segmented into five different regions. Each event will be added to one of the five regions in the queue but are only removed from the active region.
  1. **Active events:** This event occurs at current simulation time and can be processed in any order.
  2. **Inactive events:** This event occurs at current simulation time but shall be processed after all active events are processed.
  3. **Non-blocking assign update event:** This event is evaluated during some previous simulation time, but shall be assigned at this simulation time after all active and inactive events are processed.
  4. **Monitor event:** This event is processed after all active events, inactive event and non-blocking assign update event are processed.
  5. **Future event:** This event occurs at some future simulation time. Future events are divided into future inactive event and future non-blocking assign update event.
- Processing of all active events is called simulation cycle.

5.3 SENSITIVITY LIST

5.3.1 Description

- In sensitivity list controls, when all statements in the always block will started to be evaluated.
- In Verilog-1995 the signals are separated by the keyword or.
- In Verilog-2001 the signals may be separated by a comma. This new comma-separated sensitivity list does not add new functionality.
- It does make the Verilog syntax more intuitive, and more consistent with other signal lists in Verilog.
- The sensitivity list must include all input signals used by an always block to properly model combinational logic. It is easy to inadvertently omit an input signal from the sensitivity list, which can lead to simulation and synthesis mismatches.

- Verilog-2001 adds a new wild card token, which represents a combinational logic sensitivity list. The token adds to the sensitivity list all nets and variables that are read by the statements in the always block.
- Only static signal names, for which reading is permitted, may appear in the sensitivity list of a process, i.e. no function calls are allowed in the list.

5.3.2 Definition

- ***The sensitivity list is a compact way of specifying the set of signals, events on which may resume a process.***
- A sensitivity list is specified right after the keyword process. It is equivalent to the wait on statement, which is the last statement of the process statement, i.e. no function calls are allowed in the list.
- The sensitivity list is where you list all the signals that you want to cause the code in the process the code in the process to be evaluated whenever it changes the state.

5.3.3 Simplified Syntax

- The simplified syntax of sensitivity list is given by,
  (signal_name, signal_name, . . .)

5.3.4 Examples

```
DFF : process (CLK,RST)
begin
  if RST = '1'
    then Q <= '0';
    elsif (CLK'event) and (CLK = '1')
      then Q <= D;
  end if;
end process DFF;
-- DFF : process
-- begin
-- if RST = '1'
-- then Q <= '0';
-- elsif (CLK'event) and (CLK = '1')
-- then Q <= D;
-- end if;
-- wait on RST, CLK;
-- end process DFF;
```

- Here, the process is sensitive to the RST and CLK signals, i.e. an event on any of these signals will cause the process to resume. This process is equivalent to the one described in the comment section.
- A process with a sensitivity list may not contain any explicit wait statement. Also, if such a process statement is a parent of a procedure, then that procedure may not contain a wait statement as well.

5.3.5 Sensitivity List versus Wait Statement

- A process that has a sensitivity list corresponds to the some process without the list and with a wait ON statement that waits and the same sensitivity list.
- The wait ON statement must be the last one of the process statement part.
- It is illegal to have a process with sensitivity list and wait statement in the same time. Thus, a wait statement is illegal for a process with a sensitivity list.
- Process cannot have both a wait statement and a sensitivity list.

5.3.6 Advantages

1. In-depth analysis
2. Decision making
3. Quality check
4. Proper allocation of resources

5.3.7 Disadvantages

1. Based on assumptions
2. Not relative in nature
3. Not accurate

5.3.8 Applications

1. Environmental
2. Business
3. Social sciences
4. Chemistry
5. Engineering
6. In meta-analysis
7. Multi-criteria decision
8. Time-critical decision making
9. Model calibration and improvement

4.4 ZERO MODELING

5.4.1 Introduction

- The idea of signal modeling is to represent the signal via some model parameters. Signal modeling is used for signal compression, predication, reconstruction and understanding.
- Two generic model classes will be considered:
  1. ARMA, AR, and MA models,
  2. Low-rank models.

5.4.2 Concept

- A theory for lossless pole-zero modeling of speech signals is established for the description of a nasal sound.
- This theory is based on a generalized vocal tract tube model which consists of the main vocal tract, the oral tract, and the nasal tract. A pole-zero type transfer function, which turns out to be a generalized version of the existing all-pole type transfer function, is derived. Fundamental properties of the generalized vocal tract model are investigated, employing the concept of discrete-time reactance.
- A procedure for evaluating the reflection coefficients for the model is outlined. The assumption of losslessness in the modeling leads to the following two properties.
- First, the combination of two lattice structures representing the oral and the nasal tracts form one larger lattice structure when viewed at their joint point called the branch boundary.
- Second, the oral and the nasal tract render respective discrete-time reactances whose convex combination generates the discrete-time reactance at the branch boundary.
- The first property enables the combined reactance to be computed, and the second helps separate it into its components.

5.5 COMPARISON OF SOFTWARE AND HARDWARE DESCRIPTION LANGUAGES

Table 5.1

Software Language	Hardware Description Language
1. In a software language all assignments are sequential. That means the order in which the statements appear is significant because they are executed in that way.	1. HDL is the standard text based expression of temperal behaviour of the electronic circuits that describe the circuit operation, its design and test to verify its operation by means of stimulation.
2. The software language is a programming language just as C.	2. HDL look like a software programming language such as a C language.
3. Software language is a middle level language.	3. HDL is a combination of high level language and assembly level language.
4. Software language handles only sequential instructions.	4. VHL allows to handle both sequential and concurrent execution.
5. Software language cannot be used to describe and implementing hardware circuits.	5. HDL can be used for describing and implementing the hardware circuits.

Contd...

6. Software languages can be written successfully with pure logical or algorithms thinking.	6. HDL can be written successfully by considering throughout knowledge of hardware circuit.
7. Software language (C) can be usually run on powerful processor with high speed due to its high memory.	7. VHDL cannot be run on powerful processor and high speed due to its low memory and logic elements.
8. It has no difficulty to implement image processing algorithms.	8. It has great difficulty to implement image processing algorithms.
9. Software language does not include notation of time.	9. HDL includes notation of time.
10. We get different results, when order of sequence is changed.	10. HDL is always concurrent.

5.6 DELAY MODELS IN VHDL

5.6.1 Introduction

- In VHDL, the designer has the possibility to perform a signal assignment after certain amount of time, implementing the delays in the assignment.

- The delay mechanism allows introducing propagation times of described systems.

- Delays are specified in signal assignment statement. It is not allowed to specify delays in variable statements.

- Delay mechanism can be applied to signals only, which are applied to systems. Delays are not synthesizable.

5.6.2 Definition

- *Delay is mechanism allowing introducing timing parameters of specified systems.*

5.6.3 Types of Delays

- There are basically two delay mechanisms available in VHDL. They are as under :
    1. Inertial delay (default),
    2. Transport delay.

5.6.4 Inertial Delay

- If the delay mechanism in VHDL is not specified, then by default it is inertial delay.

- The inertial delay is defined using the reserved word **inertial** and is used to model the devices, which are inherently inertial. In practice this means, that impulses shorter than specified switching time are not transmitted.

- The inertial delay specification may contain a **reject** clause. This clause can be used to specify the minimum impulse width that will be propagated, regardless of the switching time specified.

- As inertial delay is the default delay, the reserved word inertial can be omitted and you may simply call it as delay.

- The inertial delay models the delay introduced by an analog port, which means, it is analogous to the delay in devices that respond only if the signal value persists on their inputs for a given amount of time.

- The inertial delay model is the default delay implemented in VHDL because it's behavior is very similar to the delay of the device. The delay assignment syntax is as under :

    b <= a **after 20 ns;**

- In this example *b* take the value of *a after* 20 ns second of inertial delay. This means that if a value varies faster than 20 ns *b* remain *unchanged*. This simulation should clarify the concept.

- At simulation start *a* and *b* are 0; a change from 0 to 1, and *b* change its value after *20 ns*; then *a* changes value going to *0* and then to *1* in *10 ns*. This delay is *less than* inertial delay of 20 ns. So *b* remains *unchanged*.
- Thus inertial delay is often found in switching circuits. It is the default delay in digital circuits. This delay is often used to filter unwanted spikes and transients on the signals.
- The inertial delay represents the time for which an input must be stable before the value is allowed to propagate too the output. In addition, the value appears at the output after the specified dealy.

5.6.5 Transport Delay

- The transport delay is defined using the reserved word **'transport'** and is characteristic for transmission lines.
- Transport delay models the delays in hardware that do not exhibit any inertial delay. This delay represents pure propagation delay, i.e, any changes on an input is from ported to the output, no matter how small, after the specified delay.
- To use a transported delay model, the keyword transport must be used in a signal assignment statement.
- New signal value is assigned with specified delay independently from the width of the impulse in waveform, i.e. the signal is propagated through the line.
- The transport delay is not the default delay implemented in VHDL and *must be specified*. This delay model is useful to describe delay line, PCB delay, wire delay. The delay assignment syntax is:

> **b <= transport a after** 20 ns;

- In this example *b* take the value of *a* after 20 ns second of *transport delay*. This means that no matter how fast *a* changes his value, *b* will follow the behavior of *a* after the amount of time specified in the delay statement. This simulation should clarify the concept.
- The simulation is the same as the inertial delay. In this case the value of *b* follow the value of *a* after 20 ns even when a has a glitch of 10 ns.
- VHDL transport and inertial delay model allow the designer to model different type of behavior on VHDL hardware implementation.
- The inertial and transport delays are very useful in test bench modeling, such as RAM, ROM, and peripheral.

5.6.6 Delta Delay

- Delta delay is a very small delay, which does not correspond to any real delay and actual simulation time does not advance.
- This delay models hardware where a minimal amount of time in needed for a change to occur, for example, in performing zero delay simulation.
- Delta delay allows for ordering of events that occur at the same simulation time during a simulation.
- Each unit of simulation time can be considered to be composed of an infinite number of delta delays. Therefore, an event always occurs at a real simulation time plus an integral multiple of delta delays.
- In VHDL simulation, all signal assignments occur with some infinitesimal delay, known as delta delay. Technically, delta delay is of no measurable unit, but from a hardware design perspective one should think of delta delay as being the smallest time unit one could measure, such as femtosecond.
- A delta or delta cycle is essentially an infintesimal, but quantized, unit of time. The delta delay mechanism is used to provide a minimum delay in a signal assignment statement, so that the situation cycle described earlier can operate correctly when signal assignment statements do not include extalicity specified delays.
- All active processes can execute in the same simulation cycle. Each active process will suspend at wait statement.

- When all processes are suspended, simulation is advanced the minimum time necessary to that some signals each take on their new values.
- Processes the determine, if the new signal values satisfy the conditions to process from the wait statement at which they are suspended.

5.7 SIMULATORS

5.7.1 Introduction

- Computer program (such as a **game** or animated flow chart) or a **dedicated device** that models (simulates) some aspects of a real life situation (such as flying an aircraft) and can be manipulated to observe the outcomes of different **assumptions** or actions, without exposing the experimenter to any **danger** or **risk**.
- A simulator has the following four things:
  1. Description design
  2. Simulator to derive the design
  3. Sometimes designs are self simulation and do not need external stimulus.
  4. Most of the times, the VHDL designers use VHDL, testbench of the kind or another to arrive the design being tested.

5.7.2 Definition

- **A simulator is a device that enables the operator to reproduce or represent under test conditions phenomena likely to occur in actual performance.**
- A simulate is a test bench maker that needs two inputs.

5.7.3 Usage Examples

1. At the science museum the students use the **weather simulator** to learn about different storm patterns in areas of the world.
2. The training of airline pilots is often accomplished by the use of an **aircraft simulator** to train them under realistic flight conditions.

5.7.4 Types

- There are two types of simulators as under :
  1. Event-based simulator.
  2. Cycle-based simulator.
- They are as discussed below.

I.　The two-phase simulation cycle :

  1. Go through all functions. Compute the next value to appear on the output using current input values and store it in a local data area (a value table inside the function).
  2. Go through all functions. Transfer the new value from the local table inside to the data area holding the values of the outputs (=inputs to the next circuit).

1.　Event-based simulator :

- Event driven signal keeps track of any change in the signal in the event queue.
- The simulator starts simulation as soon as any signal in event list changes its value.
- For this the simulator has to keep record of all the scheduled events in future. This causes a large memory overload but gives high accuracy for asynchronous design. It simulates events only.
- Gates whose inputs have events are called active and are placed in activity list.
- The simulation proceeds by removing a gate from the activity list. The process of evaluation stops when the activity list becomes empty.

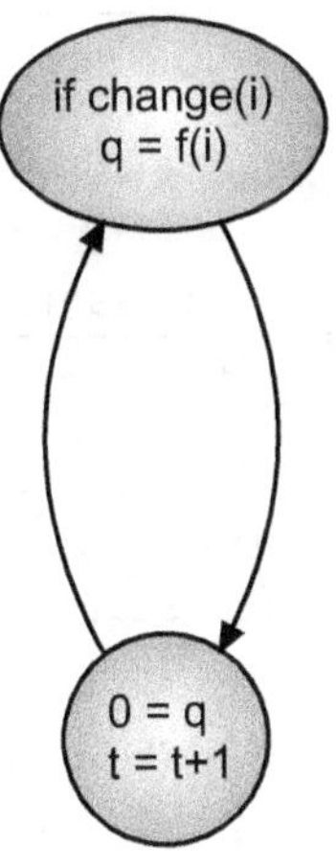

Go through all functions using current inputs and compute next output.

Update outputs and increase time with 1 delay unit

Fig. 5.2

2. Cycle-based simulator :

- Cycle-based simulation ignores intra-cycle state transitions, i.e. they check the status of target signals periodically irrespective of any events. This can boost performance by 10 to 50 times compared to traditional event-driven simulators.
- Cycle-based technology offers greater memory efficiency and faster simulation run-time than traditional pure event-based simulators.
- Cycle-based simulators work best with synchronous design but give less timing accuracy with asynchronous design.
- Signals are treated as variables. Functions such as AND, OR etc. are directly converted to program statements.
- Signal level functions such as memory blocks, adders, multipliers etc. are modeled as subroutines.
- For every input vector, the code is repeatedly executed until all variables have attained steady value.
- Compiled code simulator is efficient when used for high-level design verification. Inefficiency is incurred by the evaluation of the design when only few inputs are changing.

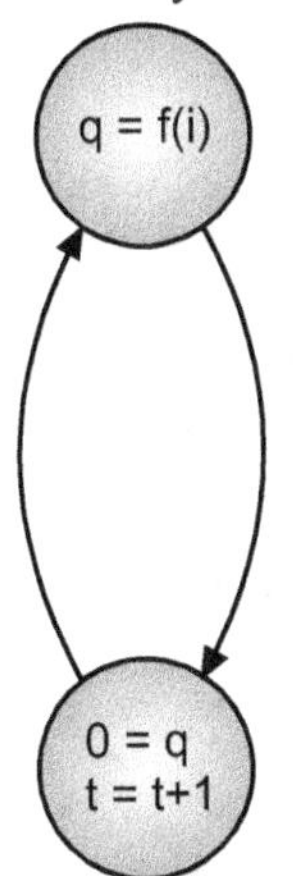

Go through all functions whose inputs has changed and compute next output.

Update outputs and increase time with 1 delay unit.

Fig. 5.3

5.8 COMPARISON OF EVENT-BASED AND CYCLE-BASED SIMULATORS

Table 5.2

Event-based Simulator	Cycle-based Simulator
1. It evaluates input looking for state change.	1. It evaluates entire design every clock cycle.
2. It schedules events in time.	2. No event scheduling.
3. It calculates time delay.	3. No delay calculation or timing checks.

Contd...

4. It stores state values and time information.	4 No such storage.
5. It is slow and has less efficient memory usage.	5. It is very fast and very efficient memory usage.
6. It identifies timing violations.	6. It does not identify timing violations.
7. It does not require a static timing analysis tool.	7. It requires a static timing analysis tool.

5.9 SYNTHESIS IN VHDL

5.9.1 Introduction

- Synthesis process may be done by human or a computer program. There is a surge of incentives to program automatic synthesis programs for VLSI designs.
  1. VLSI complexity has increased tremendously.
  2. VLSI technology has become mature and a work hours for many applications.
  3. Demand for shorter and shorter design cycle.
  4. Need to expose bigger design space across levels of abstraction.

5.9.2 Definition

- **Synthesis is the translation process from a description of a hardware device at higher abstraction level into an optimized implementation on a lower level abstraction.**
- Synthesis is an automatic method of conversion of higher level abstraction into the lower level of abstraction. It is a means of converting VHDL model into real-word hardware.

5.9.3 Steps

- To convert RTL (Register Transfer Logic) description to gates, there are three steps as under :
  1. Translation.
  2. Boolean optimization.
  3. Optimization.

1. Translation :

- RTL description is translated to unoptimized Boolean description usually with primitive gates like AND and OR gates, and flip-flops and latches.
- RTL description to Boolean equivalent circuit is not controlled by user.

2. Boolean optimization :

- Algorithms are executed on the Boolean optimized description.

3. Optimization :

- Optimization process takes an unoptimization Boolean description.
- Optimization uses number of algorithms to convert unoptimized Boolean descriptions to a very low level description (PLC FORMAT).
- Thus, when we optimize 'PLA Format', we get description. We have to reduce the logic generated by sharing common terms.
- Optimized Boolean equivalent description is mapped to actual logic gates by making use of technology library of target process.

5.9.4 Types

- There are two general categories of the synthesis process as under:
  1. Behavior-to-structure layout,
  2. Structure-to-physical layout
- We can also view the synthesis process in three different levels as under:
  1. Behavior synthesis,
  2. Logic synthesis,
  3. Physical synthesis.

Logic Synthesis

- Logic synthesis involves both combinational and sequential logic design. If a stable diagram or table is given, it goes through the state minimization process and then the state encoding process.
- After the encoding is completed truth tables or Boolean expressions can be used to describe the relationships between input, current state and output, and next state.
- The next step involves the process of technology mapping, which converts the optimized truth table or boolean expression into gates of a particular technology.

5.9.5 Expectation

- All synthesis tools do not perform all expectations and do not have all features.
- Following are some of the expectations from a good synthesis tool.
    1. Synthesis tool should perform technology specific optimizations. e.g. vendor specific FPGAs and CPLDs.
    2. Best optimization techniques should be available.
    3. Designer should have control over the synthesis process. Synthesis process can control through coding style and constraints.
    4. Synthesis tool should have broad language coverage.
    5. Synthesis tool should provide a user-friendly debugging environment.
    6. Synthesis tool should have fast compile time. Compiled time should be proportional to design density.
    7. Synthesis tool should provide a clean and seamless link to the backend tools (simulator) i.e. transition should be very easy from front to backend tools.

5.9.6 General Remarks

- The main properties of the synthesizable subset of VHDL are as under:
1. Only a single architecture for each entity to be synthesized is allowed. A second architecture presented to the system will result in the first one to be ignored. Configurations do not make sense because no confusion between multiple architectures is possible.
2. The architecture of an entity can either be a behavioral one or a structural one composed of instantiations of other entities. So, hierarchical descriptions can be used. Multiple entities per file are allowed.
3. Behavioral descriptions of an entity will have one or more processes in the architecture body. It is a good custom to separate combinational and sequential logic into separate processes.
4. Synthesizable VHDL should not contain references to absolute time such as in assignments with the after keyword. If they do, they are ignored. Signals can be delayed, but only by passing them through clocked registers.
5. Although the synthesizer can deal with many data types, it is strongly recommended to exclusively use the std logic and std logic vector data types for the I/O signals of the top level entities. These are namely the data types used in the VHDL descriptions of the synthesized circuits. Sticking to them facilitates the reuse of test benches.

5.9.7 Advantages

- The advantages of using synthesis are as under :
    1. It frees the designer from technology dependent issues.
    2. It allows the technology independent coding.
    3. The designer can guide the synthesizer to optimize his design for speed, power or area. Depending on these constraints, the synthesis will produce different netlist to optimize on selected parameters.
    4. It forces higher level abstraction.
    5. It leads to ease in designing and code portability.
    6. It enables designer to focus on larger design gates, as it is easier to relate RTL to hardware.

5.10 HDL DESIGN FLOW FOR SYNTHESIS

5.10.1 Introduction

- The main problem with logic synthesis in the original HDL-based FPGA flows that they were derived from ASIC flows. The tools worked at the primitive logic gate level, and produced gate level net lists, which needed to be mapped, packed and P and R by the FPGA vendor.

- In 1994, synthesis tools became FPGA architecture aware and could perform mapping, and some level of packing to produce a LUT/CLB-level netlist. The advantage is that synthesis tools had a better understanding of timing and area utilization.

- Around 2000, the concept of physically aware synthesis began to take hold in the FPGA world, to address problems with obtaining timing closure. Here, the product of the synthesis engine was a mapped, packed and placed CLB-level netlist. The FPGA P&R tools started with the initial placement and performed local (fine-grained) placement optimizations and detailed routing.

- Flow to HDL tools and methods convert flow-based system design into a HDL such as VHDL or verilog. Typically, this is a method of creating designs for FPGA and ASIC prototyping and DSP design. Flow-based system design is well suited to field programmable gate array design as it is easier to specify the innate parallelism of the architect.

5.10.2 Digital System Design Flow

- Design flows operate at multiple level of abstraction.
- Need a uniform description to translate between levels.
- Increasing costs of design and fabrication necessitate greater reliance on automation via CAD tools.
- Fig. 5.4 shows the block diagram of a digital system design flow.

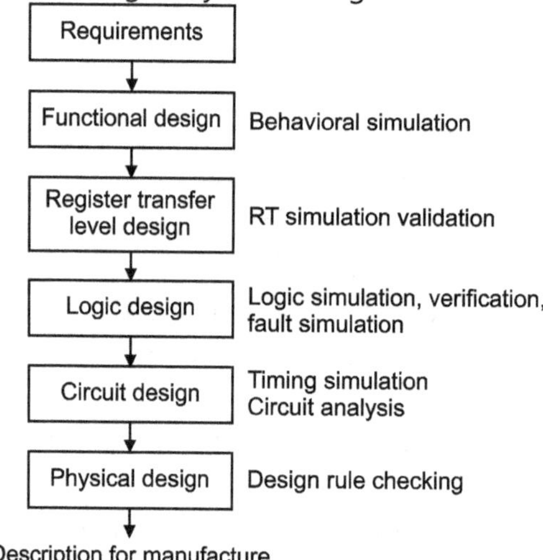

Fig. 5.4 : Digital system design flow

5.10.3 Synthesis Design Flow

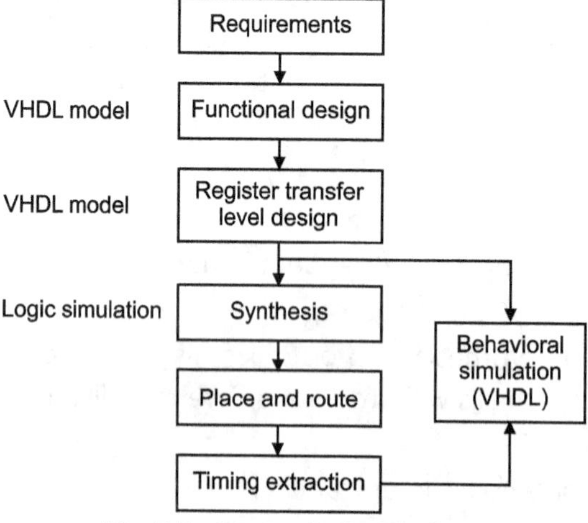

Fig. 5.5 : Synthesis design flow

- Fig. 5.5 shows the block diagram of a synthesis design flow.
- Automation of design refinement steps.
- Feedback for accurate simulation.
- Example target ASICs, FPGAs.

5.10.4 Role of HDL

- The role of HDL in synthesis is shown in Fig. 5.6.
- Design is structured around a hierarchy of representations.
- HDLs can describe distinct aspects of a design at multiple levels of abstraction.
- Interoperability: models at multiple levels of abstraction.
- Technology independence: portable model.
- Design to re-use and rapid prototyping.

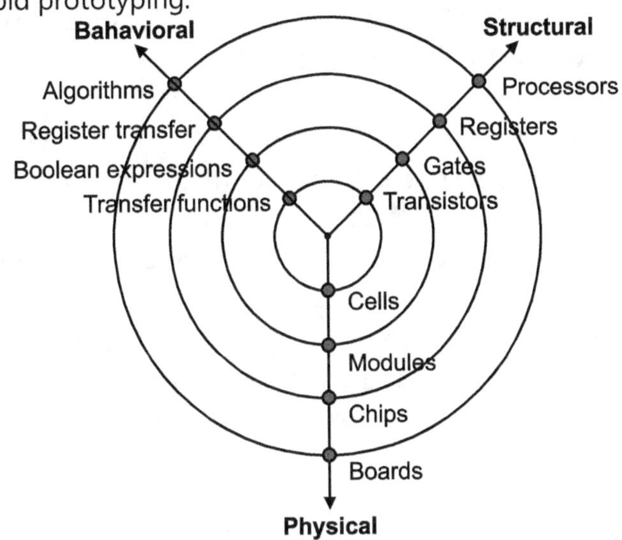

Fig. 5.6 : Role of HDL

5.10.5 Applications

1. Supercomputer architecture
2. Bioinformatics
3. Financing processing
4. Oil and gas survey data analysis
5. Embedded systems
6. System on a chip
7. CFD

5.11 EFFICIENT CODING STYLES

5.11.1 Introduction

- Effective coding standard forms on techniques that highlight problems and make bugs stand-out and visible to everyone. They are as under:
  1. They are short, simple and concise rules. They *do not attempt to cover and processify everything* and leave plenty of room for developers to exercise their own creativity.
  2. strike the right balance between formatting issues and issues that *"make the wrong code look wrong"*.
  3. They are black and white instead of vague suggestions or recommendations.
- Coding standards, when used for the right reasons and in the right manner, offer many benefits.
- Effective coding standards focus on techniques that highlight problems and make bugs stand-out and visible to everyone.

5.11.2 Definition

- **A coding style is a set of rule that a programmer user for choosing an expensive form to use in a given situation.**
- Usually, these rules are aesthetic, but sometimes, there are efficiency issues involved. A few choices that have non-obvious efficiency consequences.

- **Coding standards are set of guidelines, best practices, programming styles and conventions that develops adhere to when writing source code for a project.**
- Efficient coding style are a set of informal rules that the software development community has learned overtime which can help to improve the quality of software.

5.11.3 Few Guidelines

- Few guidelines from Linux Kernel coding rule are as under :
    1. Tabs are 8 characters, and this indentations are also 8 characters.
    2. The limit on the length of lines is 80 columns and this is a strongly professes limit.
    3. The preferred form for allocating a zeroed array is the following:
       p = kcalloc(n, sizeof(...), ...);
- Both forms check for overflow on the allocation size n* size of (- - -), and return NDLL if that occurred.

5.11.4 Few Examples of Good Coding Standards

    1. Not more than one statement per line.
    2. Line length should not exceed 80 or 100 characters.
    3. Test class must start with the name of the class they are testing followed by 'Test'.
       e.g. ServerConfigurationTest.
    4. One character variable names should only be used in loops or for temporary variables.

5.11.5 HDL Style for Performance

- A coding style is a set of rules that a programmer uses for choosing an expressive form to use in a given situation. Usually these rules are aesthetic, but sometimes there are efficiency issues involved. A few choices that have non-obvious efficiency consequences are describes here.

HDL Style for Performance :

Rule 1: Use Optimized Standard Libraries :

- Customers report up to a 3x performance increase when switching from unoptimized to optimized VHDL libraries. For ModelSim, all of the most frequently used VHDL libraries have been specifically tuned for maximum performance within ModelSim. These optimizations can be disabled by using special switches at compilation or by explicitly mapping in alternate libraries. However, the most common reason for mistaken use of unoptimized libraries. This occurs if the build environment compiles standard library source code from a non-Model Tech source. Source code for standard libraries is often included with synthesis tools or ASIC vendor libraries, and is often compiled by mistake. These unoptimized libraries will take precedent over the default ones.

Table 5.3 : ModelSim Optimization of VHDL Standard Libraries

Optimized
IEEE
std_logic_1164
std_logic_arith
std_logic_signed
std_logic_unsigned
numeric_bit
numeric_std
arithmetic
VITAL
vital_primitives
vital_timing
Std
std.standard
std.textio

- The Performance Analyzer can quickly show you when you are using an unoptimized library. If the performance report implicates a line within a library (outside of user code) then the library has not been optimized. Optimized libraries do not show up in the performance analyzer report. If the library indicated is one in the optimized list of Table 5.2, then review the steps taken to compile the design.

Rule 2: Reduce Process Sensitivity :

Avoid inefficient processes like this one:

```
inefficient : process (A, B)
begin
procedure_1(A);
procedure_2(B);
end process inefficient;
```

- Notice that every time B changes, a call is needlessly made to procedure_1. Similarly, events on A will force the redundant evaluation procedure_2. Note that if you have shared data between the two processes, you may have difficulty accurately synthesizing the correct behavior. In the example above, the
- Performance Analyzer is likely to identify excess time spent in this process.
- Two separate processes, each with the correct sensitivity list is the more efficient coding style:

```
efficient_1 : process (A)
begin
procedure_1(A);
end process efficient_1;
efficient_2 : process (B)
begin
procedure_2(B);
end process efficient_2;
```

- This is a trivial example, but processes like these appear often in the customer examples. Unnecessarily sensitive can severely impact performance.
- Also, use caution when creating processes sensitive to signals of record type. The record may contain more information than the process strictly needs, but any change to any element of the record will force a reevaluation of the process.

Rule 3: Reducing waits

- It is a common practice to use a for loop around a wait on clock to allow a specific amount of time to pass.
- This fragment delays 100 clock cycles:

```
for i in 1 to 100 loop
wait until Clk'Event and Clk = '1';
end loop;
next statement . . .
```

- While this loop is not complicated, the Performance Analyzer may identify the "wait" line as a bottleneck.
- The reason for this is the proliferation of processes waiting for signal events, even though the action taken by each process is minimal.
- Although slightly more obscure, the following fragment accomplishes the same behavior:

```
wait for (CLOCK_PERIOD_T * 100 - 1 ns);
wait until Clk'Event and Clk = '1';
next statement . . .
```

- The first fragment schedules 100 process evaluations, while the second requires only two. The behavior is the same, but the performance consequence is minimized. The final wait until Clk is needed to ensure proper synchronization with the clock signal. Without it, the "next statement" is in an unpredictable race condition with whatever is generating the clock.

Rule 4: Reduce or Delay Calculations

- The following fragment repeats the same 64-bit calculations at each evaluation of the process:

```
driver : process (Clk)

begin

if (Clk'event and Clk = '1') then

. . .

D <= Next_D_val after (CLOCK_PERIOD_T - SETUP_T);
LD <= Next_LD_val after (CLOCK_PERIOD_T - SETUP_T);

. . .
```

- The drive times are repeatedly calculated. With the simple use of a constant, two 64-bit operations per clock cycle are removed:

```
driver : process (Clk)

constant DRIVE_T : time := (CLOCK_PERIOD_T - SETUP_T);

begin

if (Clk'event and Clk = '1') then

. . .

D <= Next_D_val after (DRIVE_T);
LD <= Next_LD_val after (DRIVE_T);

. . .
```

- Another good rule of thumb is to delay calculations until they are needed. Here is an example of an inefficient call to a function:

```
Wrong . . .

val := to_integer(array);
if (A = B) then
Result <= val;
else
Result <= "00000000";
End if;

. . .

Right

. . .

if (A = B) then
val := to_integer(array);
Result <= val;
Else
Result <= "00000000";
End if;

. . .
```

- The example on the left makes the "to_integer" call every evaluation, whether the result is used or not.

Rule 5: Integers versus Vectors

- Arithmetic operations on Standard Logic Vectors (SLVs) are expensive compared to integer operations.
- Consider converting an SLV to an integer, performing the operations and converting the integer back to an SLV. Integer conversion costs are small compared to costs of even simple SLV operations. In the example below the unsigned vector "value" is used in a simple comparison (> 0) and a subtraction.

```
. . .
if (value > 0) then -- <-- Slow
value <= value - 1; -- <-- Slow
else
value <= startValue;
end if;
. . .
```

- The performance analyzer might identify the two lines as being the slowest part of this process. Suppose that for the purposes of your design, two states would suffice for "value". You could then use an integer instead:

```
int_value := to_integer(value);
if (int_value > 0) then -- <-- Fast
int_value := int_value - 1; -- <-- Fast
else
int_value := to_integer(startValue);
end if;
value <= to_unsigned(int_value, 8);
. . .
```

- The performance of the process would be significantly improved. If you have testbench code that generates only two-state or four-state behavior, it should be relatively straight-forward to write the testbench using integers instead of std_logic_vectors.
- For maximum performance, use ranged integers in entity declarations instead of std_logic_vectors. With both the interface and internal state represented in integers, the simulator will be able to process the design much more efficiently. This is a fairly dramatic step, and you should make sure that your synthesis tools can properly handle ranged integers in your design.

Rule 6: Avoid Slicing Signals

- If a signal is sliced, vector optimizations cannot be applied.

```
signal A_sig : std_logic_vector (63 downto 0);
. . .
A_sig(3) <= '1';
```

- The signal is probably used in several places in the design. Even a single bit slice propogates an unoptimized vector to all affected processes. The introduction of a temporary variable can give you the functionality of a bit slice, without the performance penalty:

```
Wrong                              . . .
for i in 0 to 4 loop
A_sig(i) <= mask_calc(i);
end loop
. . .
Right
```

```
for i in 0 to 4 loop
tmp_A(i) <= mask_calc(i);
end loop
A_sig <= tmp_A;

. . .
```

- In the example A_sig is kept whole, while the bit slicing occurs for the temporary variable "tmp_A". The costs of slicing the temporary variable and the additional assignment are small in comparison to the penalty of an unoptimized signal vector.

Rule 7: Optimize Everything over 1%

- The ModelSim Performance Analyzer will identify the lines of code that consume the greatest CPU time and display these lines in ranked order in the performance profile window. Double clicking a line in the report will bring up the source file window with the file and line displayed.

5.12 OPTIMIZING ARITHMETIC EXPRESSIONS

5.12.1 Introduction

- In 1939, a linear programming of simulation of a problem that is equivalent to the general linear programming problem was given by the soviet economist *Leonid Kantorovich*, who also proposed a method for solving it.

- The linear programming problem was first shown to be solvable in polynomial time by *Leonid Khachiyan* in 1979, but a larger theoretical and practical breakthrough in the field came in 1984 when *Narendra Karmarkar* introduced a new *interior-point method* for solving linear-programming problems.

- Arithmetic expressions are on alternative approach to description of logic circuits sharing useful properties of *Reed-Muller expression* and permitting at the same time. Simplified representations of multi-output functions.

- In many applications, where Boolean functions need to be analyzed, arithmetic expressions provided a better in sight into related problems and offer efficient solutions. Applications of arithmetic expressions in logic design dates back to early 1950s, and a renewed interest in this subject is to ever increasing challenges of probabilities verification of logic circuits and requests for compact decision diagram and related representations, where other methods do not provide acceptable solutions or cannot be used for large space and time requirements.

- Arithmetic expressions are closely related to *Reed-Muller expressions*, since they are defined in terms of the same basis, however, with variable and function values interpreted as integers 'o' and 'I' instead of logic values.

- In this way, arithmetic expressions can be considered as integers counter parts of *Reed-Muller expressions*. Due to this interpretation, many efficient algorithms for *Reed-Muller expressions* as well as optimization methods can be extended to arithmetic expressions. In particular, as for Reed-Muller expressions, the optimization of arithmetic expressions in the number of non-zero coefficients count can be performed by choosing literals of different polarity for integer counterparts of switching variables in terms of which arithmetic expressions are defined.

5.12.2 Definition

- Any n-variable switching function f given by the truth-vector $F = [f_0, f_1, f_2, ..., f_{2^n-1}]$ can be represented by the positive polarity arithmetic expression (PPAE) defined as

$$f(X_1, ..., f_n) = X(n) A(n)F \qquad ... (5.1)$$

$$\text{where,} \qquad X(n) = \overset{n}{\underset{i=1}{\otimes}} [1 \; x_i], \qquad ... (5.2)$$

$$A(n) = \overset{n}{\underset{i=1}{\otimes}} \begin{bmatrix} 1 & 0 \\ -1 & 1 \end{bmatrix}, \qquad ... (5.3)$$

where $\otimes$ denotes the **Kronecker product**, and addition and multiplication are arithmetic operations. $A(n)$ represents the arithmetic transform matrix of order n.

If each variable can appear as complemented or uncomplemented, but not both, the related expressions are fixed-polarity arithmetic expressions (FPAEs) given as

$$f(X_1, \ldots, f_n) = X_H(n)\, A_H(n)F \qquad \ldots (5.4)$$

where,

$$X_H(n) = \bigotimes_{i=1}^{n} [1\ x_i^{h_i}], \qquad x_i^{h_i} = \begin{cases} x_i & , h_i = 0 \\ \overline{x_i} & , h_i = 1 \end{cases} \qquad \ldots (5.5)$$

$$A_H(n) = \bigotimes_{i=1}^{n} A^{h_i}(1) \qquad \ldots (5.6)$$

$$A^{h_i}(1) = \begin{cases} \begin{bmatrix} 1 & 0 \\ -1 & 1 \end{bmatrix} , & h_i = 0, \\[6pt] \begin{bmatrix} 0 & 1 \\ 1 & -1 \end{bmatrix} , & h_i = 1. \end{cases}$$

- Therefore, FPAEs are uniquely characterized by the polarity vectors $H = [h_1, \ldots, h_n]^T$, $h_i \in \{0, 1\}$, where $A_i = 1$ shows that the i-th variable is complemented and written as $\overline{X_i}$.

- An FPAE can be given by the FPAE spectrum. A_f^h is calculated as

$$A_f^h = A_H(n)F.$$

5.12.3 Optimization Algorithm

- The optimization algorithm is a modification of the two-term **iterative common** subexpression elimination algorithm. The only difference between the delay aware optimization algorithm and the original algorithm is that in the delay aware version, the true value of each divisor instead of just calculating the number of additions that can be selected without increasing the delay beyond the maximum specified delay. The delay for each instance of each divisor is calculated using the value is the only different between the delay aware optimization algorithm and original algorithm, only that procedure is illustrated below:

Find_true_value(d, MaxDelay)

d = distinct divisor in set of divisors {D}

{ d_instances} = Set of instances of divisor d

{P_i} = Set of expressions containing
       any of the instances in { d_instances }

{ Allowed_ instances} = Φ;
     for each expression P_i in {P_i}

{

{Valid_d}_i = Set of non-overlapping instances
         that can be extracted in Pi without
         making its delay > Max Delay;

{Allowed instances}
     = {Allowed_instances} ∪ (Valid_d)_i ;

}

True value = |{Allowed_instances}| ;

return True_value;

}

- This procedure is called for each candidate divisor d, with the given limit on the maximum delay Max-Delay. In this procedure, first the set of all expressions, which contain instances of that divisor are collected in the set {P_i}.

- Now for each expressions, P_i in that set, the number of non-overlapping instances of divisor of that do not increase the critical path of P_i beyond max Delay is calculated.

- The true value of the divisor is then equal to the sum of such instances over all the expressions in {P_i}.

5.12.4 Arithmetic Optimization Forms

- Design Compiler uses the properties of arithmetic operators (such as the associative and commutative properties of addition) to rearrange an expression so that it results in an optimized implementation. You can also use arithmetic properties to control the choice of implementation for an expression. Three forms of arithmetic optimization are discussed in this section: merging cascaded adders with a carry, arranging expression trees for minimum delay, and sharing common sub expressions.

Merging Cascaded Adders with a Carry

- If your design has two cascaded adders, and one has a bit input, VHDL Compiler replaces the two adders with a simple adder that has a carry input. Example below shows two expressions where cin is a bit variable connected to a carry input. Each expression results in the same implementation.

  Example: Cascaded Adders with Carry Input

  z <= a + b + cin;

  t <= a + b;

  z <= t + cin;

  t <= a + cin;

  z <= t + b;

Arranging Expression Trees for Minimum Delay

- If your goal is to speed up your design, arithmetic optimization can minimize the delay through an expression tree by rearranging the sequence of the operations. Consider the following statement

  Example: Simple Arithmetic Expression

  Z <= A + B + C + D;

- The parser performs each addition in order, as though parentheses were placed around the expression as follows.

  Z <= ((A + B) + C) + D;

Considering Signal Arrival Times

- To determine the minimum delay through an expression tree, Design Compiler considers the arrival times of each signal in the expression. If the arrival times of each signal are the same, the length of the critical path of the expression in

  Example above equals three adder delays. The critical path delay can be reduced to two adder delays if you add parentheses to the first statement as shown.

  Z <= (A + B) + (C + D);

Using Parentheses

- You can use parentheses in expressions to exercise more control over the way expression trees are constructed. Parentheses are regarded as user directives that force an expression tree to use the groupings inside the parentheses. The expression tree cannot be rearranged in such a way that it violates these groupings. If you are not sure about the best expression tree for an arithmetic expression, leave the expression ungrouped. Design Compiler can reconstruct the expression for minimum delay.

- To illustrate the effect of parentheses on the construction of an expression tree, consider the following example.

  Example: Parentheses in an arithmetic expression

  Q <= ((A + (B + C)) + D + E) + F;

5.13 SHARING OF COMPLEX OPERATOR

5.13.1 Introduction

- The complex operator may be mobile telecommunication services which have been shown impressive uptake in the past decade. In particular, in developing countries, mobile telephony has played a vital role in making *cellular services* available to a part of the population that did not have access to such services previously.

- However, considerable advances are required to increase the penetration of mobile services and to improve competition in the cellular market, in particular in rural areas in developing countries. The roll-out of mobile networks requires high sunk investments and the need to recover those by charging the user heavily for accessing mobile services.

- This often makes mobile services less affordable and may discourage operators to innovate and migrate to new technologies in emerging markets. It may also cause licensed mobile network operators (MNO) to obstruct the entry of new operators in the market and additionally, it may be too costly for new entrant operators to rollout mobile networks in rural and less populated areas, resulting in exclusion of a part of the population or certain regions from access to mobile telecommunication services.

- The traditional mobile network operation strategy is characterized by a high degree of vertical integration where the MNO acquires and develops the sites needed for rolling out the network, plans the network architecture and topology, operates and maintains the network and customer relationships, creates, markets and provides services to its end users.

- However, technology migration, such as the introduction of third generation (3G) and 3.5G wireless technologies on top of 2G networks, and the introduction of 4G technologies including LTE, is becoming increasingly rapid and complex. Regulatory requirements also mandate coverage of areas that is not attractive from a business perspective. With growing competitive intensity and rapid price declines, mobile operators are facing increased margin pressure and the need to systematically improve their cost position.

- In the emerging market context, both in urban and rural areas infrastructure sharing should be adopted as an imperative for sustained telecommunication growth.

5.13.2 Infrastructure Sharing

- Mobile infrastructure sharing may also stimulate the migration to new technologies and the deployment of mobile broadband, which is increasingly seen as a viable means of making broadband services accessible for a larger part of the world population.

- Mobile sharing may also enhance competition between mobile operators and service providers, at least where certain safeguards are used, without which concerns of anticompetitive behaviour could arise.

- Ultimately, mobile network sharing can play an important role in increasing access to information and communication technologies (ICTs), generating economic growth, improving quality of life and helping developing and developed countries to meet the objectives established by the World Summit on the Information Society (WSIS) and the Millennium Development Goals established by the United Nations.

- Different forms of infrastructure sharing are possible, ranging from basic unbundling and national roaming, to advanced forms like collocation and spectrum sharing. In the MENA (Middle East and North Africa) region, National roaming is used extensively in countries.

- The infrastructure sharing is the most cost-efficient design principle for any new roll-out in emerging markets and the best approach for technology migration and consolidation [12], the cost savings potential from infrastructure sharing are earned through sacrificing some of the control that the standalone operator has over its network, thereby impacting the ability of operators to compete and differentiate themselves based on network quality.

5.13.3 Technical Approach

- Network sharing is a very complex process. There are a variety of options that may be considered when assessing the viability of infrastructure sharing. Those options range from the sharing of towers and other infrastructure facilities to sharing an entire mobile network. This section identifies a number of technical options, dividing them into three basic categories:

(i)   passive sharing,

(ii) active sharing and

(iii) roaming-based sharing.

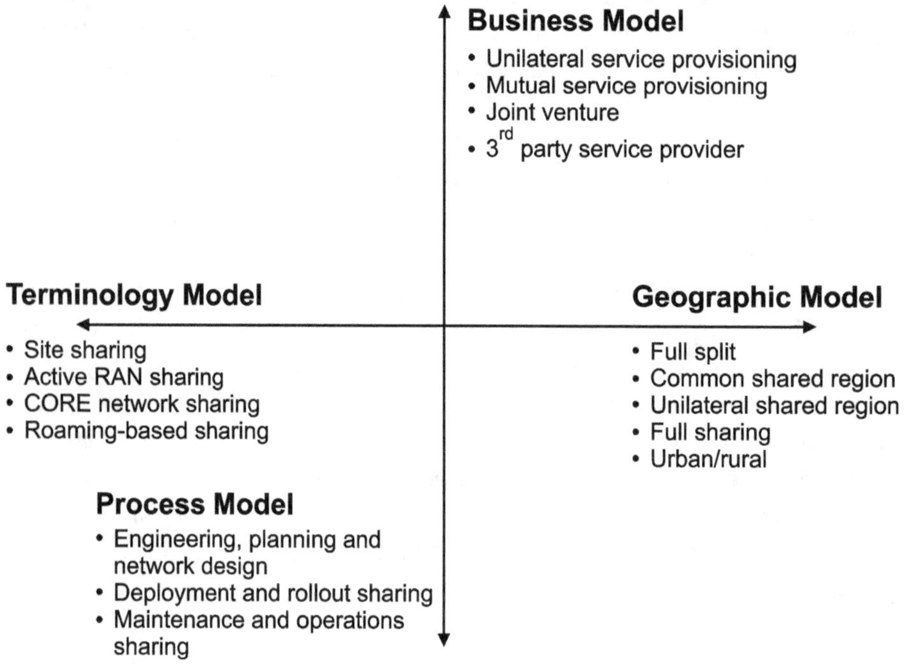

Business Model
- Unilateral service provisioning
- Mutual service provisioning
- Joint venture
- 3rd party service provider

Terminology Model
- Site sharing
- Active RAN sharing
- CORE network sharing
- Roaming-based sharing

Geographic Model
- Full split
- Common shared region
- Unilateral shared region
- Full sharing
- Urban/rural

Process Model
- Engineering, planning and network design
- Deployment and rollout sharing
- Maintenance and operations sharing

Fig. 5.7 : Dynamics of infrastructure sharing

1.   Passive Infrastructure Sharing:

- Passive infrastructure sharing is identified as the options available for mobile operators intending to share passive elements in their radio access network. Such sharing of radio sites, termed 'site sharing' or 'collocation' has become popular since the year 2000.

- Normally, the operators enter an agreement to share sites directly, but lately, there are enabling third parties involved in such agreements, which provides towers to the telecommunication operators. These so called 'tower companies' already have established footprint in mature markets where as they are coming up in the emerging markets as well, like India and the MENA region.

- Generally, site sharing involves sharing of costs related to trading, leasing, acquisition of property items, contracts and technical facilities and the sharing of passive RAN infrastructure, i.e.,

    1.   Masts and Pylons, electrical or fiber optic cables;

    2.   Physical space on the ground, towers, roof tops and other premises,

    3.   Power supply, air conditioning, alarm installations and other passive equipments;

    4.   Protecting access;

- Site sharing allows operators to reduce both capital (CAPEX) and operating (OPEX) expenditure by reducing their investments in passive network infrastructure and in network operating costs. Site acquisition costs and expenses for civil works account for up to 40% of the initial investment to the fixed assets.

- Passive infrastructure sharing may be an effective option for upgrading second generation (2G) mobile services to third generation (3G) mobile communications and broadband wireless access technologies in emerging markets.

2.   Active Infrastructure Sharing:

- The active sharing of facilities is an advanced technical model which involves a mutual sharing of not only passive, but also 'active elements' of the network that can be managed by the operators - installed in base stations and mobile network equipment, access node switches and finally the management systems of fiber optic networks.

- Active infrastructure sharing is more complex, since it covers the essential elements of value creation in the chain of economic activity. Many countries regulate the sharing of active infrastructure, fearing that the practice will promote anticompetitive behavior, such as agreements on price or service offerings.
- There are three types :
  1. Antenna sharing
  2. Base Station sharing
  3. Sharing of Base Station Controller
  4. Core Network sharing

5.13.4 Roaming

- National / International roaming is a form of sharing allowing customers of a mobile network operator to use mobile services when they are in an area not covered by its operator. From the beginning of 2G networks, roaming has always been employed as a means of virtually extending the geographic coverage of an operator by allowing its subscribers to use another operator's network.
- International roaming is the natural solution to serve one's customers abroad, where the operator has no license and no business. Roaming is also used on a domestic basis, as national roaming, typically to grant to a new entrant – or "greenfield" –operator nationwide coverage right from the start, when the operator rolls out its network initially in the urban and suburban areas and is not yet present in the rural areas.
- Roaming-based options in the context of network sharing, instead, mean that one operator relies on another operator's coverage for a certain, defined footprint on a permanent basis. Such dependence can be either unilateral or bilateral, regionally split or for the network as a whole.
- A geographically separated network does not contain shared nodes. Each operator has its own carrier and its own MNC and builds its own access networks and core network and covers different areas of the country.
- There are three options for providing national coverage. The first option is to sign a national roaming agreement, and to share the load. If the operators decide to retain dedicated independent core networks or only share the radio access network in a certain region, the "shared RAN with gateway core" solution can be deployed.
- Similar from a point of view of addressable cost items, compared to the active RAN sharing solutions outlined above, this approach, however, does not require specific features in the RAN equipment, as the sharing is fully implemented by roaming features that need to be implemented in the core network.
- The shared RAN is connected to the core networks of the sharing partners via a so-called gateway core consisting of MSC, SGSN, and visitor location register (VLR). In this solution, either frequencies are pooled, or only the frequency spectrum of one of the participating operators is used, such that there is no independent control of the traffic quality and capacity for the operators.
- If only one spectrum is used, capacity is substantially reduced; the pooling of frequencies is again subject to restrictive regulatory policies. Another solution is consistent with current releases in 3GPP, where both operators have their own access network, but a shared core network. In this case, the operators only retain that portion of the core network separate which also an MVNO (mobile virtual network operator) would own, i.e. home location register (HLR), authentication and billing system.

5.13.5 Technical Constraints

- The sharing of network infrastructure requires coordination and cooperation between the involved network operators, with the increase in the level of sharing. Such cooperation shall bring forth multiple constraints on the activities of the concerned operators, which ultimately limits their flexibility of operation.
- These constraints particularly affect the operational elements in the deployment and operation of networks and can have an impact on the ability of operators to differentiate themselves in terms of services or quality of services.
- The following items are presented for analysing the technical constraints that would apply to the operators based on the expected level of sharing.

1. **Related to Passive Sharing:**
   (a) The qualifying sites to share (electromagnetic compatibility, models blankets, site area, optimizing 2G3G).
   (b) Installation of equipment on the shared site (access and safety, engineering site deployment schedule of– operators).
   (c) The operation and maintenance of equipment (on-site, monitoring and steering of networking– equipment).

2. **Related to Active Sharing:**

A. **Sharing of antennas**
   (a) The need for common choices, affecting the quality of service (technical diversity reception and transmission, radio planning, architecture of the antenna, use of Tower Mast Head Amplifier (TMA).
   (b) The influence on the planning of the radio antenna amplifier linearity over several frequency bands.
   (c) Taking into account in planning radio 3 dB loss induced by the coupling of the common antenna, for– the separation of equipment connected to it.

B. **Sharing NodeB**
   (d) The use of NodeB containing at least two carriers (a significant difference between frequency bands of operators provides additional technical complexity).
   (e) Limited number of operators (typically 3 or 4).
   (f) A risk of lead single manufacturer solutions (in particular because of the interoperability links NodeB – RNC).
   (g) Potential conflicts on the quality levels depending on the services available (power sharing).
   (h) The operation and maintenance of shared assets.

C. **Sharing the RNC**
   (i) Management of separation of the RNC functions (radio access configuration, performance management and quality of radio services).
   (j) Interoperability between equipment from different manufacturers (hardware and software– configuration).
   (k) Interoperability between RNC and shared RNC, connected through the IuR interface [6] to ensure the handover (soft handover).

D. **Core network sharing**
   (l) A choice of design of equipment common (NodeB, RNC, MSC, SGSN) to handle the traffic associated with the provision of services of each operator.
   (m) A design package from core network management and service quality.
   (n) The need to support intelligent network protocols consistent to ensure continuity of customer service of each operator when roaming on the shared network.

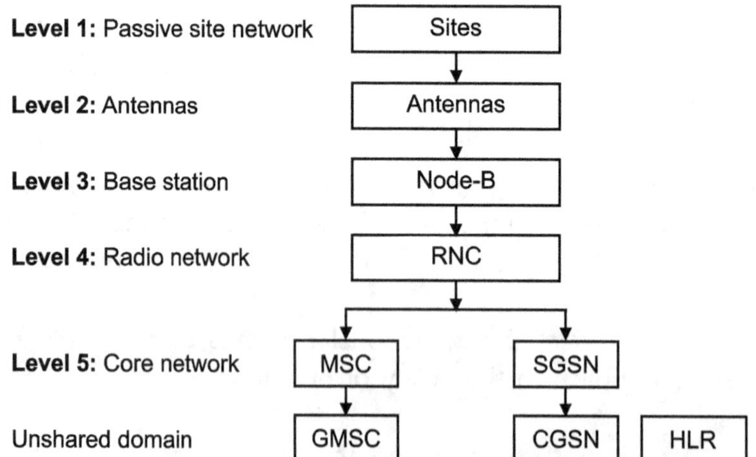

Fig. 5.8 : Different Levels of infrastructures sharing in mobile networks

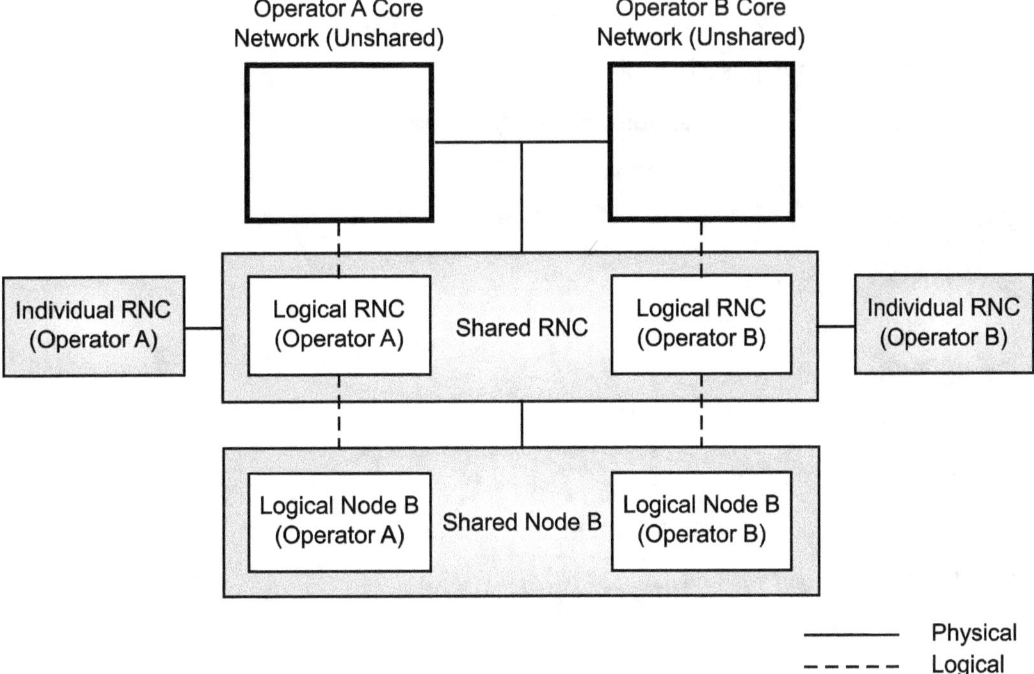

Fig. 5.9 : RNC Sharing

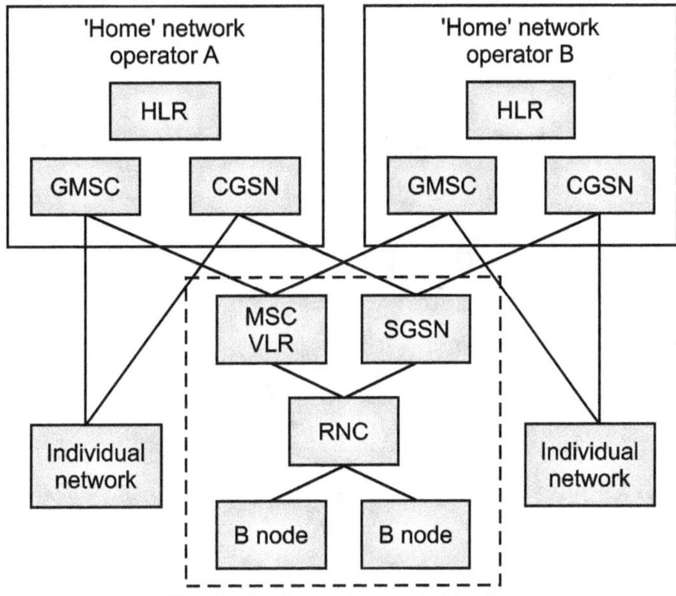

Fig. 5.10 : Core Network Sharing

Practice Questions

1. Define event scheduling.

2. What is the significance of sensitivity list ?

3. What is the difference between HDL language and software language ?

4. Differentiate between inertial delay and transport delay.

5. Explain delta delay with proper example.

6. Define synthesis.

7. Explain different ways to achieve efficient coding in VHDL.

8. Explain optimizing arithmetic expression with example.

9.　Explain sharing of complex operator with example.

10.　What do you mean by event-based simulator ?

11.　What do you mean by cycle-based simulator ?

12.　Explain the significance of using parentheses.

13.　What is simulation ?

14.　What are the types of simulation ?

15.　What do you mean by simulator ?

16.　What are steps of simulator in working ?

17.　How simulation is executed ?

18.　Explain the simulation processes.

19.　Compare functional and timing simulations.

20.　Compare software language with HDL.

21.　What do you mean by constraints ?

22.　What are ways to achieve constraints in simulation of VHDL ?

23.　Describe simulation with block diagram.

24.　Define the following terms : (a) Event, (b) Sensitivity list, (c) Delta delays.

25.　What are the types of simulators ?

26.　Explain event-based and cycle-based simulators.

27.　Compare event-based and cycle-based simulators.

28.　What do you mean by synthesis ?

29.　What are the different inputs and outputs of synthesis tool ?

30.　Why VHDL is a synthesizer and not a simulator ?

31.　What is RTL ? What are its advantages while writing VHDL code ?

32.　Write a short note on synthesis issue.

33.　Draw and explain synthesis flow.

34.　Explain optimization process in synthesis.

35.　Explain step by step process of synthesis.

36.　What are three steps of synthesis ?

37.　What are the advantages of using synthesis ?

38.　What are the expectations from synthesis ?